Mastering Arduino Programming

A Quick Guide to Arduino Programming with Simple Do it yourself projects

DANIEL

STONES

Copyright

Table of Contents

Introduction

What Is Arduino?

Arduino is an open-source electronics creation platform, which is based on free hardware and software. It is flexible and easy to use for creators and developers. This platform allows the creation of different types of single-board microcomputers that can be used by the developer community for different types of use.

To understand this concept, you are first going to have to understand the concepts of free hardware and free software. Free hardware is the device whose specifications and diagrams are publicly accessible so that anyone can replicate them. This means that Arduino offers the bases so that any other person or company can create their boards, being able to be different between them but equally functional when starting from the same base.

Free software is computer programs whose code is accessible by anyone so that anyone who wants can use and modify it. Arduino offers the Arduino IDE (Integrated Development Environment) platform, which is a programming environment with which anyone can create applications for Arduino boards so that they can be given all kinds of utilities.

The project was born in 2003, when several students from the Institute of Interactive Design of Ivrea, Italy, tried to facilitate the access and use of electronics and programming. They did it so that electronics students would have a cheaper alternative to the popular BASIC Stamps, plates that were worth more than a hundred dollars at the time, and that not everyone could afford it.

The result was Arduino, a board with all the necessary elements to connect peripherals to the inputs and outputs of a microcontroller and which can be programmed in Windows, macOS, and GNU / Linux. A project that promotes 'learning by doing' philosophy, which means that the best way to learn, is tinkering.

The Arduino is a board that is almost completely dependent on an ATMEL microcontroller. Microcontrollers are integrated circuits in which instructions can be recorded, which you write with the programming language that you can use in the Arduino IDE environment. These instructions allow you to create programs that interact with the circuitry on the board.

The Arduino microcontroller has what is called an input interface, which is a connection in which we can connect different types of peripherals on the board. The information from these peripherals that you connect will be transferred to the microcontroller, which will be in charge of processing the data that comes through them.

The type of peripherals that you can use to send data to the microcontroller largely depends on what use you are considering. They can be cameras to obtain images, keyboards to enter data or different types of sensors.

It also has an output interface, which is responsible for carrying the information that has been processed in the Arduino to other peripherals. These peripherals can be screens or speakers, which reproduce the processed data, but they can also be other boards or controllers.

Arduino is a project and not a specific board model, which means that by sharing its basic design you, can find different types of boards. They come in various shapes, sizes, and colors to meet the needs of the project you are working on. There are simple ones or ones with improved features, Arduinos oriented to the Internet of Things or 3D printing and, of course, depending on these characteristics you will find all kinds of prices.

Also, Arduino boards have other types of components called Shields or backpacks. It is a kind of board that is connected to the mainboard to add a myriad of functions, such as GPS, real-time clocks, radio connectivity, LCD touch screens, development boards, and a very long etcetera of elements. There are even stores with sections specialized in these elements.

History of the Arduino project

The Arduino project is the result of a team of teachers and students from the Interaction Design School of Ivrea (Italy). They encountered a major problem during this period (between 2003 - 2004). The tools necessary for the creation of interactivity projects were complex and expensive, somewhere between 80 and 100 euros. These often too high costs made it difficult for the students to develop many projects and this slowed down the concrete implementation of their learning.

Until then, prototyping tools were mainly dedicated to engineering, robotics, and technical fields. They are powerful but their development processes are long and they are difficult to learn and use for artists, interaction designers, and, more generally, for beginners. Their concern then focused on making hardware cheaper and easier to use. They wanted to create an environment close to Processing. This programming language was developed in 2001 by Casey Reas and Ben Fry, two former students of John Maeda at MIT, himself initiator of the DBN project.

In 2003, Hernando Barragan, for his graduation thesis, undertook the development of an electronic board called Wiring, accompanied by a free and open programming environment. For this work, Hernando Barragan reused the sources of the Processing project. Based on an easy-to-access programming language adapted to the development of design projects, the Wiring card, therefore, inspired the Arduino project (2005).

As with Wiring, the objective was to arrive at a device that was simple to use, with low costs. The codes and plans are "free" (that is to say, the sources of which are open and can be modified, improved, and distributed by the users themselves) and "multi-platform" (independent of the operating system used).

Designed by a team of professors and students (David Mellis, Tom Igoe, Gianluca Martino, David Cuartielles, Massimo Banzi as well as Nicholas Zambetti), the Arduino environment is particularly suited to artistic production as well as the development of designs that can find their achievements in industrial production. The name Arduino originates from the name of the bar the team used to hang out at. Arduino is also the name of an Italian king, a historical figure of the city "Arduind'Ivrée", or even a masculine Italian first name which means "strong friend".

The "Interaction Design Institute Ivrea" (IDII) school is today located in Copenhagen under the name "Copenhagen Institute of Interaction Design". Casey Reas was himself a teacher at IVREA, for Processing, at the time.

Why Is The Use Of Arduino Popular?

A software and hardware platform for creating digital objects, Arduino allows you to program electronic circuits that interact with the environment around them. Connected in particular to sound, thermal and motion sensors, these inexpensive electronic circuits, called micro-controllers, can in return generate images, actuate an articulated arm, send messages on the Internet, etc. Tens of thousands of artists, designers, engineers, researchers, teachers, and even companies use it to realize incredible projects in multiple fields:

1. **Rapid prototyping of innovative projects using electronics**, Arduino facilitating experimentation upstream of the industrialization phase;

2. **Artisanal production of digital objects and low- cost machine tools** in the perspective of a culture of technological appropriation favoring DIY and resourcefulness;

3. Collection and analysis of scientific data (environment, energy, etc.) for educational, research or citizen appropriation purposes;

4. **Live show**, thanks to the many interactive functions offered by Arduino, it is possible to create VJing performances, use the movement of dancers to generate real-time sound and visual effects in a show;

5. **Digital art installations,** Arduino allowing the creation of works of art interacting autonomously with the public;

6. **Fashion and textile design,** several stylists and designers investing in this creative field by exploiting the possibilities offered by the integration of electronics, in particular in clothing (e-textile);

7. **Educational projects for students**, professionals, or the general public depending on the initiators of these initiatives: colleges, specialized training centers, or Media Labs.

Advantages and Disadvantages of Arduino

The expression "Arduino " encompasses the world of electronics and programming to offer you a creative world in a DIY fashion. It is an electronic card which is a cross between a very simplified computer and a programmable controller. A ready-to-use card that can be programmed to control anything you want.

An Arduino generally has:

1. A microcontroller to store your program called "sketch" and run it

2. A USB port to interact with the card through your computer to load your "sketch" into the

microcontroller also called the upload. The USB port is also a means of powering the card

3.   7-12v power supply to power up the card if you do not want to go through the USB

4.   Pins delivering a voltage to power other electronic components in 3.3v or 5v

5.   Analog/digital pins to connect a whole bunch of sensors/components (temperature sensor, sound, ultrasound, brightness, LCD, LED or touch screen).

Arduino was designed by electronics and programming enthusiasts with an educational purpose for people new to electronics to realize and prototype your projects simply and quickly. Learning is very fast to prototype with Arduino because you only have to acquire 2 things: the basics in electronics and programming. The basics are very quick to learn, around 1-2 hours depending on your skills and you will have designed your first sketch and prototype.

Here is a non-exhaustive list of possible projects to do with Arduino:

1.   make robots

2.   manage cameras

3.   control motors

4.   water your plants after some time

5.   distribute kibble if your dog's bowl is empty,

6.   know the temperature of your rooms

7.   turn on or off a lamp according to a presence

8.   automate a mailbox

9.   retrieve consumption information via EDF remote computing

10.   Make your alarm

Arduino boards can be found alone or sometimes sold in a pack like the "Inventor" kit we had already mentioned here, which includes a whole bunch of components allowing you to quickly start various projects, just to familiarize yourself with this area.

CHAPTER ONE

Advantages

1. Arduino is "Open Source". This means that you can take the original drawing, modify it, and use it to produce the map and sell it without paying copyright fees. "Open Source" has made it possible to quickly distribute Arduino boards around the world to form a huge community that improves and designs new boards that are always more efficient.

2. The price. As the scheme is free and you do not pay any duty on its use, industries have taken the opportunity to produce the various cards. Some respect the official diagram, as well as the components, originally recommended which gives a price of around 20-25 Euros in Europe for the most popular model, the UNO. Others produce it by using lower quality components which allows them to lower the cost of the card enormously (it is possible to have it for less than 10 Euros by looking well). They are called clones. In terms of use, the clones and the official ones are very similar whatever it is necessary with certain clones to carry out certain technical manipulations to be able to program them, but the main difference lies in the quality of the card. A little advice: if you want to start with Arduino, invest in an official rather than a clone. It will save you a bad experience and a waste of your time. You can take clones later when you master the map.

3. The community. A community is very important in this kind of project. This facilitates exchanges between users on different cards. There are many forums and documentation online to help you use the map and overcome any issues you may encounter.

4. Simplicity. Arduino rhymes with simplicity. The project was designed so that beginners in electronics and programming can design prototypes very quickly of what they have in mind. In a few hours of learning, you will be able to design your prototype.

5. The multiplatform. To program an Arduino board and make it do what you have in mind, you have to connect it to a computer and use the Arduino IDE, the software allowing you to program all the Arduino boards. The IDE is cross-platform by being available on Windows, Mac OSX, and Linux.

6. The "shields". These are additional cards that connect directly and easily to an Arduino card to increase its possibilities by adding for example a GPS, an Ethernet or Wifi interface, an LCD screen, a sensor, etc. Of course, it is possible to add its functionalities through components which have the advantage of being less expensive but much more tedious to use.

7. No limit. To use an Arduino is to adopt it. You will have no limits in your design projects on Arduino. Finally if only one: your imagination. For example, many 3D printers work based on Arduino. This is particularly the case with the BCN3D + which is powered by an Arduino Mega. You can do all kinds of projects with this little card.

Projects you can do using Arduino

A home automation project in Arduino

Arduino can allow you to prototype and complete many projects regardless of your chosen field. Designing a connected object for home automation has several advantages that the Arduino already has.

1. The cost. Nowadays, home automation is becoming more and more democratized, but certain technologies and certain connected objects can seem quite expensive at first glance, which can slow you down in the automation of your home. Designing a connected object does not cost that much in most cases. If you want to make, for example, a presence detector with the recovery of the number of LUX (unit of measurement of luminance) to be able to light a connected bulb automatically, this will cost you less than 15 Euros.

2. The essential. In DIY mode, you focus on the essentials, what you want, and what you need and nothing else. No more superfluous.

3. The gratification. What could be more rewarding than designing something with your little hands? If also, it helps you in your home, it's even more pleasant.

4. The communication protocol. You choose how your connected object will communicate. Bluetooth, radio frequency, Wi-Fi, Ethernet, infrared. The possibilities are many and now you have the choice. Something that you do not necessarily have with connected objects already designed using proprietary protocols

5. An alternative to what is done on the home automation market. Did you know that it is possible to create your own RFXCom for around twenty euro with an Arduino board?

Disadvantages

1. Since the programming is not carried out in assembly, the price to pay for the use of the libraries is a delay in the execution of the instructions, a few microseconds that in the case of devices of everyday use are irrelevant, but significant when it comes to doing data acquisition.

2. The fact that the platform comes already assembled takes away flexibility from the projects, so for example we would be obliged to use a space and shape according to the Arduino PCB, to overcome this, you must work with a different microcontroller than the platform and design PCBs from scratch as with PICs.

3. Thirdly is the presence of Arduino clones in the marketplace. Below are the disadvantages of the Arduino clones

1. Lack of support for the community and the development of the Arduino project. By purchasing a clone, we no longer support the development of all activities performed by members of the Arduino community. These include the development of new versions of the plates or the important educational work they carry out.

2. They are made in China plates. Most of the original plates are normally made in Italy (where the project comes from) or in the United States. Therefore its sale in a way contributes to the development of the local economy. Still, any manufacturer can sell original Arduino boards. To do this, you only have to acquire the necessary licenses for the commercial use of your brand and logo. This also contributes to supporting the community. The Arduino project has long licensed brands to Chinese companies as well. In this way, we can also find original plates that come directly from China and are still original.

3. Worse finishes. Although not always, some of the clones have poorer quality finishes. For example, in one of the clones used in a Bike Pixel prototype, the metal that surrounds the entrance of the USB port bends slightly from inserting and removing the cable to connect the battery (it still works perfectly).

4. Technical problems. For example, from time to time it is not possible to send compiled code the first time. The solution is usually to use the reset button or just restart the Arduino IDE.

So you decide. But before deciding on one, do not forget to consider all the advantages and disadvantages of using original Arduino boards or clones. The ideal I suppose is a compromise and buy both original plates and clones.

Chapter Two

Types of Arduino Boards

Arduino Uno

Arduino UNO Rev3 is the last revision that exists at the moment of this board. It is a small electronic board with a programmable microcontroller on its PCB. In addition to the said chip, it also includes a series of pins as inputs and outputs that can be used by programming the chip to do different things. In this way, electronic projects can be created very easily.

This board arises from the Arduino project, an Italian project started in 2005 that focused on developing open hardware and software for students mainly. The first designs were directed at an institute in Ivrea, Italy. At that time the students of this educational center used the famous BASIC Stamps that I have already mentioned previously. These had a considerable cost, and they weren't that open.

Before all that, Hernando Barragán had created a development platform called Wiring, a project inspired by the famous Processing programming language. With this as a basis, they went to work to develop low-cost and simple tools for students. So they set about creating a hardware board with a PCB and a simple microcontroller, as well as creating an IDE (Integrated Development Environment).

As Wiring already used a board with an ATmega168 microcontroller, the following developments followed in the same direction. Massimo Banzi and David Mellis would add ATmega8 support for Wiring, which was even cheaper than version 168. And so the first germ of what is now Arduino UNO arises. The Wiring project was then renamed Arduino.

The name of the famous project originated in a bar in Ivrea, where the founders of the project met. The bar was called Bar di Re Arduino, which in turn was named after Arduino from Ivrea, king of Italy until 1014.

Given the potential of these plates, more support was added from the community to move forward and create more plates. Additionally, electronic component suppliers and manufacturers began designing specific Arduino-compatible products. As is the case with Adafruit Industries. From here arose numerous shields and additional modules for these plates.

Given the overwhelming success, the Arduino Foundation was also created, to continue promoting and grouping the efforts of the Arduino project. A similar model to other similar organizations such as the Linux Foundation, the Raspberry Pi Foundation, RISC-V Foundation, etc.

As of this moment, many variants of Arduino have been generated, with different form factors and diverse microcontrollers, as well as many accessories:

- 28BYJ-28

- Gyroscope

- ACS712

- ESP8266

- etc

Arduino UNO detailed information

This Arduino UNO board has some characteristics that make it unique, and it has a series of differences concerning other Arduino boards that we are going to highlight.

Technical characteristics, scheme, and pinout

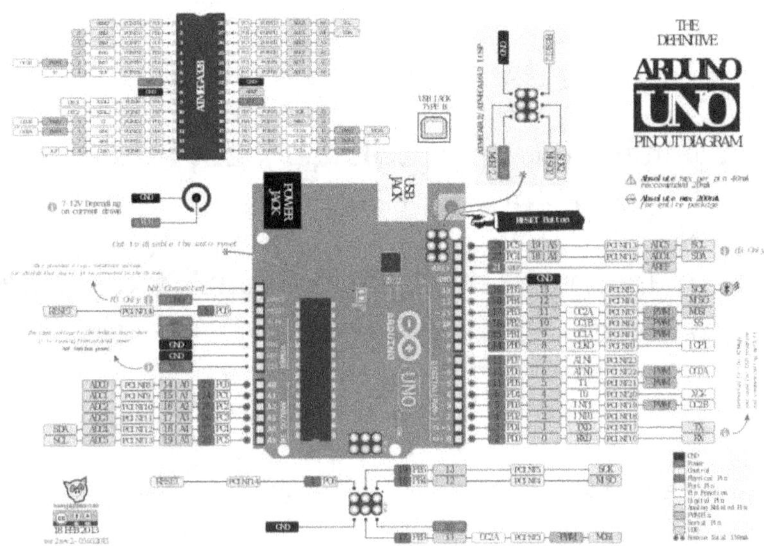

The pinout and technical characteristics of the Arduino UNO Rev3 board are important to know how to use it properly; otherwise you will not know the limits and the correct way to connect all the electronic components to their available pins and buses.

Starting first with its characteristics, you have:

1. Atmel ATmega328 microcontroller at 16 Mhz

2. Onboard SRAM memory: 2KB

3. Integrated EEPROM memory: 1 KB

4. Flash memory: 32 KB, of which 0.5 KB is used by the boot loader, so they cannot be used for other purposes.

5. Chip working voltage: 5v

6. Recommended supply voltage: 7-12v (altho-ugh it supports 6 to 20v)

7. Continuous current intensity: 40mA for I/O and 50mA for the 3.3V pin.

8. I/O pins: 14 pins, of which 6 are PWM.

9. Analog pins: 6 pins

10.   Reset button to restart the execution of the program loaded in memory.

11.   USB interface chip.

12.   Oscillator clock for signals that need rhythm.

13.   Power LED on PCB.

14.   Integrated voltage regulator.

15.   Price around € 20.

Regarding the pins and connections available on the Arduino UNO board:

1 **Barrel Jack or DC Power Jack:** it is the connector of the Arduino UNO board to power it electrically. The card can be powered by a suitable jack and by an adapter to supply 5-20 volts. If you are going to connect a large number of elements to the board, you will likely have to overcome the 7v barrier to be sufficient.

2 **USB:** the USB port is used to connect the Arduino board to the PC, in this way you can program it or receive data from it through the serial port. That is, basically it will help you load your Arduino IDE sketches into the internal memory of the microcontroller so that it can execute it. It can also fulfill the power function for the hob and the elements connected to it.

3 **VIN Pin:** You will also find a VIN pin that allows you to power the Arduino UNO board using an external power supply if you don't want to use the USB or the previous Jack.

4 **5V:** supplies a voltage of 5V. The energy that will reach it comes from one of the three previous cases by which you can power your hob.

5 **3V3:** this pin allows you to power 3.3v up to 50mA for your projects.

6 **GND:** it has 2 ground pins, to connect the ground of your electronic projects to them.

7 **Reset:** a pin to reset by sending a LOW signal through it.

8 **Serial Port:** it has two pins 0 (RX) and 1 (TX) to receive and transmit TTL serial data respectively. They are connected to the microcontroller on their USB-to-TTL pins.

9 **External Interrupts** 2 and 3, pins that can be configured to trigger interrupts with a rising, falling edge, or a high or low value.

10. **SPI:** the bus is on pins marked 10 (SS), 11 (MISOI), and 13 (SCK) with which you can communicate using the SPI library.

11. **A0-A5:** these are the analog pins.

12. **0-13:** these are the digital input or output pins that you can configure. A small integrated LED is connected to pin 13 that if this pin is high it will light up.

13. **TWI supports** TWI communication using the Wire library. You can use pin A4 or SDA and pin A5 or SCL.

14. AREF: reference voltage pint for analog inputs.

Datasheets

Being an open-source board, you will not only find the datasheet as in the case of many other electronic products. You can also download many other documents and electronic diagrams that will help you understand how this Arduino UNO board works internally and even to build your own Arduino implementation yourself. For example, you have at your disposal the following official information:

1 Datasheet of the Arduino UNO Rev3 Atmel ATmega microcontroller, to keep in mind the voltages, currents, and other characteristics that you must take into account for its operation.

2 Pinout or pin mapping.

3 EAGLE files with schematics for makers.

4 Electronic diagrams of the Arduino UNO board.

5 PCB dimensions.

Differences with other Arduino boards

Arduino UNO Rev3 is the ideal board for all those who are starting to use this type of board. Plus, there are starter kits to get started with everything you need to be included. This kit not only includes a large number of electronic components to start practicing but also a very detailed manual to help you in each step.

However, there are other versions or formats of Arduino boards that are very useful for other more advanced applications or to implement a project in which size matters. The main differences between boards are mainly in the type of integrated microcontroller, some being somewhat more powerful and with more memory to include much more sophisticated sketches or programs, and the number of pins available. But if we compare the three best-selling boards, the differences are as follows:

1 Arduino UNO Rev3: see a section with technical characteristics.

2 Arduino Mega: the price rises above € 30, with dimensions somewhat larger than the UNO board. Also, it includes a more powerful ATmega2560 microcontroller that also works at 16Mhz but has 256KB of flash memory, 4KB of EEPROM, and 8KB of SRAM for more complex programs. Plus, it also has more pins, with 54 digital I / O, 15 PWM, and 16 analogs.

3   Arduino Micro stands out for its small size, being smaller than the UNO, although of similar price. In this small space, it integrates a smaller ATmega32U4 microcontroller, but it also works at 16Mhz. The memory is equal to that of UNO, except SRAM, which has 0.5KB more. The pins have also been increased despite the small size, with 20 digital, 7 PWM, and 12 analogs. Another difference is that it uses micro-USB for its connection instead of USB. Being so small, it is not compatible with shields like the previous two.

4   Arduino Leonardo

Arduino has several boards, several flavors with which to satisfy different needs. One of the most popular development boards, along with the Arduino UNO, is the Arduino Leonardo. This board with a programmable microcontroller hides one of the most powerful features of the board line when compared to one of its sisters.

Of course, this official board of the Arduino Foundation is compatible with all the electronic components. This will give you the freedom to combine the Leonardo plate with a multitude of components to create the most varied projects you can imagine.

This Arduino Leonardo board has great similarities to the Uno, even in appearance. But you should not confuse them, since there are notable differences between the two.

Technical characteristics, scheme, and pinout

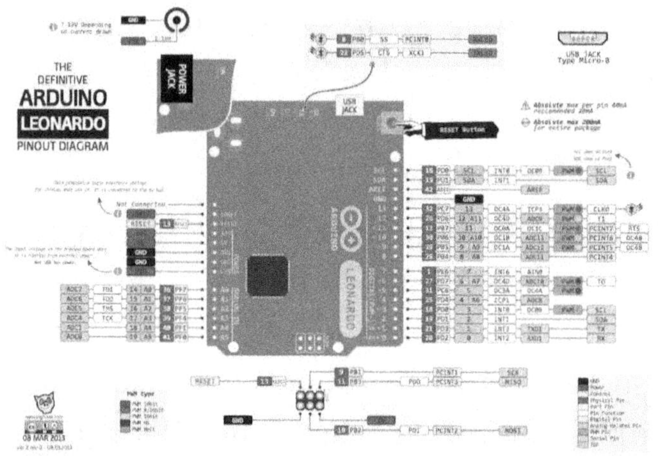

One of the main characteristics that you should know about Arduino Leonardo is its pinout, that is, the pins or connections it has. As you can see in the image above, it is not the same as the UNO Rev3 board. There are some differences between the quantity, limits, and buses.

On the other hand, you should also know its technical characteristics, which are summarized in:

- **Microcontroller**: Atmel ATmega32u4 at 16 Mhz.

- **SRAM memory**: 2.5 KB

- **EEPROM**: 1 KB

- **Flash**: 32 KB, but you have to subtract 4 KB used for the bootloader.

- **Operating voltage:** 5v

- **Input voltage (recommended) :** 7-12v

- **Input voltage (maximum limit) :** 6-20v

- **Digital** I / O pins: 20, of which 7 are PWM.

- **Analog input pins**: 12 channels.

- **Current intensity per I / O pin:** 40mA

- **Current intensity for 3.3v pin :** 50mA

- **Weight and dimensions**: 68.6 × 53.3mm and 20 grams.

- **Price**: 18 - 20 € approximately. You can buy it on Amazon.

Datasheets

As is often the case with official Arduino boards, there are a lot of diagrams, data, and documentation in this regard, even to be able to create a board derived from it as it is open-source. From the official website of the project, you will be able to find a lot of information to download about Arduino Leonardo and thus learn more about how it works. For example:

- Datasheet of the Arduino Leonardo Atmel ATmega microcontroller, to understand more about the microcontroller chip (MCU) it includes and its operation.

- EAGLE files with schematics for makers.

- Electronic diagrams of the Arduino UNO board.

Differences with other Arduino boards

The idea is to compare it with the most similar board, and that is Arduino UNO Rev3. If you compare Arduino Leonardo with UNO, you can see many similarities, but also differences that are vital if you have doubts about buying one or the other.

Physically it appears to have the same dimensions and the same number of pins. Also, they are arranged in the same way. The power supply is also the same, and even the frequency provided by the frequency generator. Also, the A0-A5 could be configured as digital with the pinMode function (pin no., Mode).

Well, one of the main differences between both development boards is in the microcontroller. While UNO is based on ATmega328, Arduino Leonardo is based on ATmega32u4 in its most recent revisions. In the case of the ATmega328, it does not have integrated USB communication, so a converter is required for that serial port. A function that the integrated circuit ATmega16u2 does.

In the case of the ATmega32u4, it does have that USB communication already implemented, so that the second chip is not necessary. That, at a practical user level, makes a difference. When you connect the Arduino UNO board, a virtual COM port is assigned for communication. While in Leonardo the plate is recognized by the computer as if it were a USB device such as a mouse or keyboard. This gives the possibility of using mouse and keyboard functions.

Of course, having another MCU also varies across some memory data. From the 32 KB flash of

Arduino UNO with 0.5 KB reserved for the bootloader, it goes to 32 KB and 4KB used by the bootloader in Leonardo. For SRAM it goes up from 2 KB to 2.5 KB and for EPROM it remains the same in both. Another difference lies in the channels of the analog inputs. Arduino UNO has only 6 channels, while in the Arduino Leonardo, there are 12 channels. That's for A0-A5, and for pins 4, 6, 8, 9, 10, and 12 which would correspond to channels A6-A11.

As for PWM, Leonardo has one more than ONE. In addition to the same ones for ONE, another is added to pin 13. The rest will be the same for both cards, that is, it will be on pins 3, 5, 6, 9, 10, and 11.

You will find more differences in I2C communication. Both can use TWI, but the difference is where the pins intended for the serial or SDA data line and the clock or SCL line are placed. In UNO they are on pins A4 and A5. But in Leonardo, you have them in 2 and 3 respectively. Slight difference, but enough that UNO's hats or shields are not fully compatible with Leonardo.

As for SPI communication, on the Arduino UNO, you have pins 10, 11, 12, and 13, for the SS, MOSI, MISO, and SCK signals respectively. This is not the case on the Leonardo, as it has a specific ICSP connector, a 6-pin male connector near one end of the card. Another reason that could make UNO shields not work for you.

For external interrupts, there are also some changes. In UNO you have two pins for it, pin 2 (interrupt 0) and pin 3 (interrupt 1). In the case of Arduino Leonardo, they extend to 5 pins. They are pins 3, 2, 0, 1, and 7 for interrupt 0, 1, 2, 3, and 4 respectively. There is also another change between the two boards that many tend to forget, and it is the type of USB cable necessary to connect both boards to the PC. While in UNO an AB cable is used, in Leonardo an A-micro B is needed.

3. Arduino Mega

If the Arduino UNO Rev3 board is too small for you and you want to create more advanced projects and enjoy more power, then what you are looking for is an Arduino Mega board, another of the available models created by the same developers as the original board, but equipped with a faster microcontroller, more memory, and more pins to program.

Arduino Mega has many similarities with Arduino UNO, but some differences make it very special for all makers looking for something more. In general, if you are just starting it is not the best choice, but it is if you have already exploited the capabilities of UNO and want to go further.

Arduino Mega is another official development board based on the Atmel ATmega2560 microcontroller, hence its name. Also, it includes 54 digital input and output pins, of which 15 can be used as PWM outputs. It also has 16 analog inputs, 4 UARTs as serial ports for hardware, a 16Mhz crystal oscillator, a USB connection, a power connector, an ICSP header, and a reset button.

As you can see, compared to the Arduino UNO, it has higher capacities, which also leads to a slight increase in its price. However, it is not expensive at all, it only costs a few Euros more and you can find it in many specialized stores. For about € 40 you have the Arduino UNO board, along with a plastic case.

It contains everything you need for your microcontroller, so you only have to worry about assembling your DIY project, connecting the board via USB to your computer, downloading the sketch you have created with the Arduino IDE, and putting it to work.

You should know that, unlike previous boards, the Arduino Mega does not use an FTDI USB-to-serial controller chip. Instead, it uses an ATmega16U2 chip in its latest revisions (Rev1 and Rev2 used the ATmega8U2). That is, it has a USB-to-serial converter programmer. This board is ideal for a multitude of advanced projects, such as serving as a brain for 3D printers, industrial CNC robots, etc. And they are fully compatible with Arduino UNO shields, so you will find a multitude of compatible elements and a large community always ready to help with your questions and problems.

And if you want to know more about compatible electronic components and modules, there are a lot of them explained step by step on the official arduino website with everything you need to put them to work. For example:

- 28BYJ-28

- Gyroscope

- ACS712

- ESP8266

- etc

Detailed information on Arduino Mega

Mega Pinout

Micro pinout

Pro Mini Pinout

The Arduino Mega board has everything you can find on the Arduino Uno Rev3 board, but with some additions that make it more powerful, as I have already mentioned.

Technical characteristics, scheme, and pinout

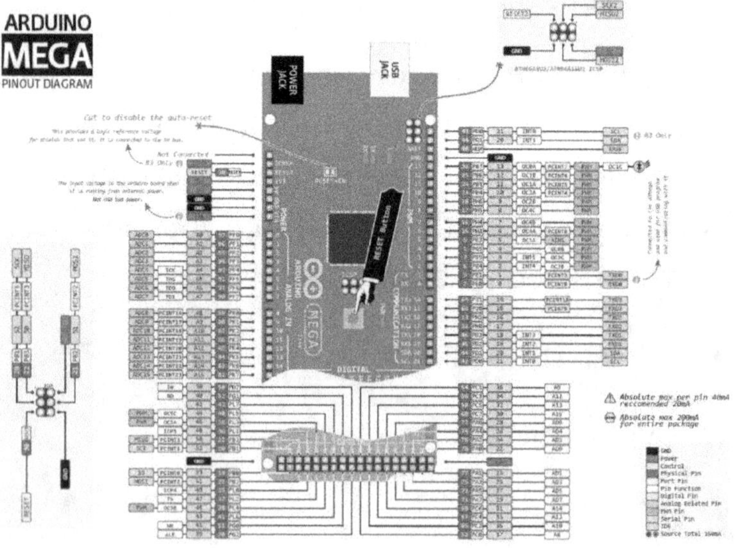

The technical characteristics of the Arduino Mega board that you should know are:

- **Atmel ATmega2560 microcontroller** at 16 Mhz

- **256 KB flash memory** (8KB used by the bootloader that cannot be used for your programs)

- **8 KB SRAM** memory.

- **4 KB EEPROM** memory.

- **5v operating voltage**

- **Input voltage** 7-12v

- **Input voltage limits:** 6-20v

- **54 digital pins,** of which 15 can be PWM. They can be configured by Arduino IDE code as inputs or outputs.

- **16 analog input pins.**

- **4 UARTs, USB**, RX, and TX pins for communication, and also TWI and SPI.

- **Power pins:** 5v to supply current to the projects as long as the board is being fed with between 7 and 12v or by the 5v USB. The 3v3 pin can supply a voltage of 3.3 volts. The GND pins can be used to ground your projects. While the IOREF pin is the pin on the board to provide the reference voltage with which the microcontroller operates.

- **The current for each I / O pin** is 40mA DC.

- **The current delivered by pin 3v3** is 50 mA.

I would also like to add that Arduino Mega has a resettable poly-fuse to protect its USB port of the computer to which you connect the board. This way you will avoid damages due to short circuits in your projects or over currents that may happen. That's an additional layer of internal protection that this version implements that kicks in if more than 500 mA is applied to the USB port, automatically breaking the connection until that overload is removed.

Datasheets

You can also download a technical sheet or datasheet with everything you need to know about the electronic details of this product, the maximum currents, and voltages allowed to avoid damaging the board, a complete pinout, and a large amount of information that you will like to have. To do this, you can download it from the official website.

Arduino Nano

Arduino Nano is another version in which you can find the famous Arduino development board. It is small, but do not be fooled by its size, it hides a lot of possibilities. It's like a real Swiss army knife. With it, you can create a multitude of projects in which it is important to keep consumption and size at bay.

Like all Arduino and compatible boards, it has similarities with other of its older sisters, although it also has certain unique and different technical characteristics from the others. In this section, you will see all those similarities and differences to understand everything you need to know about this board and start developing your DIY projects with Arduino Nano.

Arduino is already a classic in the world of free hardware and the world of makers. With its development and software beaches, you can create a multitude of projects where the limit is your imagination and well ... some technical limitations of course. But they allow you to learn electronics, programming and also create real wonders. Even professional projects are based on these development boards. In the case of the Arduino Nano, it is a stripped-down version of the Arduino UNO. This minimizes the energy demand you consume and also means less space is needed to house the bale, making it ideal for projects where size is important.

This is not exactly a miniaturized Arduino UNO board, as you will see there are some important technical differences. It is also not an alternative to LilyPad. But it shares other characteristics and the essence that is present in all Arduino projects. Of course, it can be programmed with the same Arduino IDE as the rest.

Technical characteristics

The Arduino Nano board has some technical characteristics that you should know before starting with it, in addition to evaluating if it is what you need for your project or it does not meet your expectations.

Those technical characteristics are:

- It is a small, flexible, and easy-to-use microcontroller board.

- It is based on the Atmel ATmega328p micro-controller or MCU in versions 3.x and ATmega168 in previous versions. In any case, it works at a frequency of 16 Mhz.

- The memory consists of 16 KB or 32 KB flash depending on the version (2 KB used for the boot loader), with 1 or 2 KB of SRAM memory and a 512 byte or 1 KB EEPROM depending on the MCU.

- It has a supply voltage of 5v, but the input voltage can vary from 7 to 12v.

- It has 14 digital pins, 8 analog pins, 2 reset pins, and 6 power pins (Vcc and GND). Of the analog and digital pins, they are assigned several extra functions such as pinMode and digitalWrite and analogRead for analogs. In the case of analogs, they allow a 10-bit resolution from 0 to 5v. On digitals, 22 can be used as PWM outputs.

- Does not include direct current outlet.

- It uses a standard mini USB for its connection with the computer to program or power it.

- Its power consumption is 19mA.

- The PCB size is 18x45mm with a weight of only 7 grams.

Pinout and datasheet

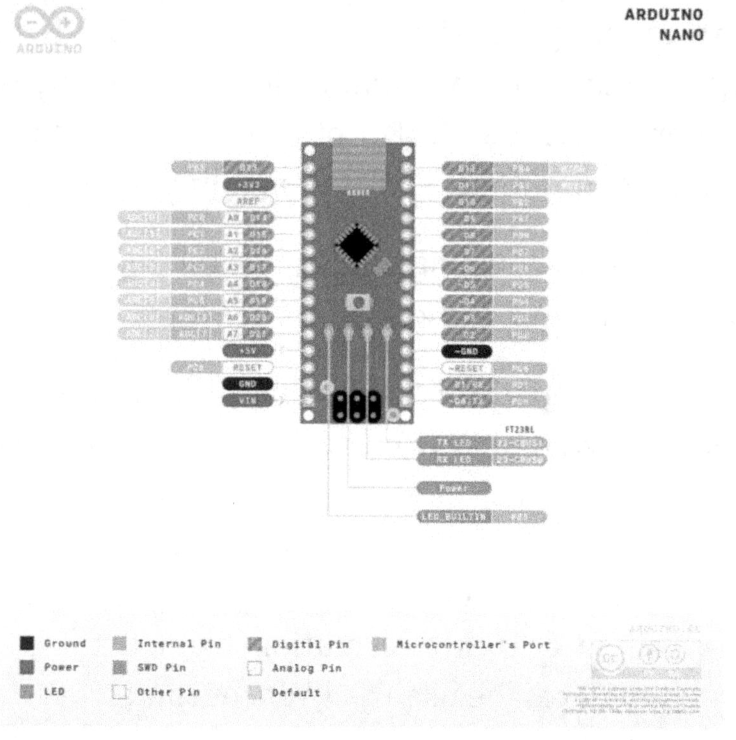

In this image courtesy of Arduino, you can see the pinout or the predisposition of pins and connections that you can find on this development board. As you can see, Arduino Nano does not have as many I/O pins as its sisters, but it does have a considerable number for most projects.

Differences with other Arduino Mini and Micro boards

Within the official Arduinos, you can find those versions that we have been talking about in this section, such as UNO, Mega, etc. One more is this Arduino Nano, which has the following differences that you have seen in the previous sections.

However, to summarize the most prominent ones, these are the most important concerning the other official reduced-size plates:

- It was designed with the same goal as the Arduino Mini, only that the Nano has a miniUSB port to program it and power it.

- Its price is between the Arduino Mini and the Arduino Micro.

Compatibility

The Arduino Nano board is compatible with all kinds of electronic components like the rest of the boards. There is no limitation of any kind beyond its maximum current and voltage limitations.

Arduino Mini

This is a board microcontroller board developed by Arduino.cc and is based on Atmega328. Performs almost the same functions with other Arduino boards, however, it is different from the Arduino Uno in terms of the design of the PCB, size, voltage regulation, and clock speed. The Arduino Uno comes with two voltage regulators, that is to say, 5V and 3.3 V, while the Arduino Pro Mini comes with a single voltage regulator. Arduino Pro mini is widely available in two versions, the 5V, and 3.3 V to operate at 16MHz and 32 MHz, respectively. However, you can get both versions as they are available separately, with a single voltage regulator in comparison with the Arduino Uno which comes with two voltage regulators, 5V and 3.3 V which runs at 16 MHz. The Arduino boards play a vital role in the development of integrated systems and other electronic projects. These plates were developed to provide an easy combination of hardware and software that give a fast way for people without technical knowledge to get practical experience with the plates. These boards come with everything you need to develop projects that have some connection with the automation.

Technical Characteristics of the Arduino Pro Mini

- Arduino Pro Mini board developed by Arduino.cc is a microcontroller board which has a microcontroller Atmega328 built in the interior of the plate.

- The Arduino Pro Mini board comes with 14 digital I/O. One can comfortably use just 6 pins to provide a PWM output. The board also features a total of 8 analog pins.

- It is very small in comparison with the Arduino Uno, that is to say, 1/6 of the total size of the Arduino Uno.

- There is only a voltage regulator built into the plate, 3.3 V or 5V depending on the version of the board.

- The Pro Mini runs at 8 MHz for the version of 3.3 V that is half of the Arduino board to One that operates at 16MHz.

- There is a USB port available on the board and also lacks a scheduler built-in.

- The labeling of the regulator defines the version of the plate, KB33 represents the edition of 3.3 V and KB50 represents the edition of 5V. However, the version of the plate also can be obtained by measuring the voltage between the GND and the Vcc pin.

- This plate does not come with connectors that are often included to offer you the flexibility of soldering the connector in as you choose to, based on the prerequisites and the available space for each project.

- Compared to other Arduino hardware boards, the Arduino Pro Mini is open-source. You can modify and use the plate according to your requirements, as all the data and support related to this motherboard are readily available.

- The ability of over current protection is another feature that makes this device safe to use in applications in which the passage of the current can affect the quality and effectiveness of your Arduino project.

- The Arduino Pro mini board features a flash memory that reaches up to 32KB. Surprisingly, it uses only 0.5 for a boot loader. The memory of the flash is used to save written codes on to the board. The flash memory is believed to be non-volatile and is able to store information even if the connection is lost with the power supply.

- SRAM is a static random access memory of 2KB. The memory RAM is of a nature highly volatile and depends mainly on the constant power supply.

- The EEPROM features up to 1KB of memory. It is basically ROM (read-only memory) that you can easily program and erase at anytime. This memory may be erased using electrical signals higher than normal.

- The software of the Arduino board is called an IDE (Integrated development environment). This software is often used to program the Arduino board. Sketch is the name of the code you write when programming the Arduino board.

- Other plates are available for purchase in the market, but the Arduino Pro Mini in particular comes with an LED built into the board. The LED blinks when you run a program or even compile programs on the board.

Arduino Pro Mini Pinout

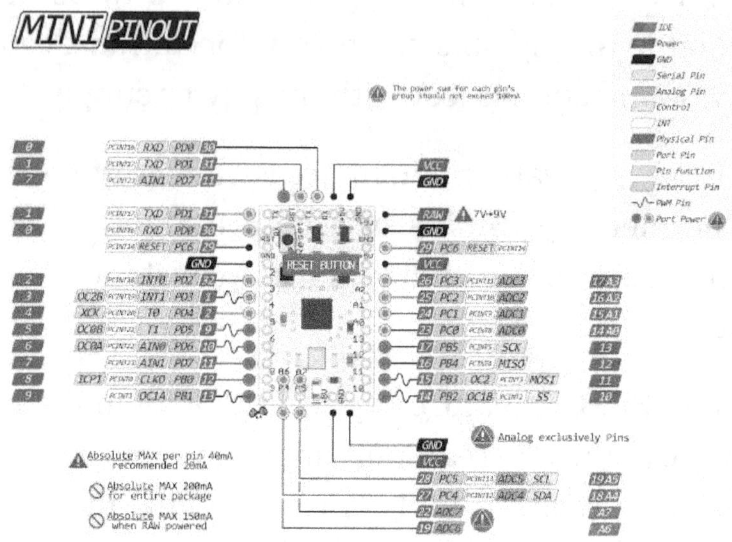

It is a compact and very small board in comparison with other Arduino boards. However, its compact and small size makes the Pro MINi board to be useful and compatible when working on most projects with Arduino. Each pinboard of the Pro Mini has its own function that is specific to it and associated with the plate.

- **GND**. There is more than one pin built on the board that can be used according to the needs when you need more pins of land for the project.

- **TXD AND RDX.**Also you can use these pins as a means of serial communication. TXD represents the transmit data in the series. RXD is used to receive the data.

- **AIN0 and AIN1.** These pins are connected to the comparator procedure.

- **VCC.** This refers to the regulated voltage which one can easily regulate to 5V or 3.3 V depending on the version of the board.

- **RAW**. This pin is used to supply the voltage to the crude oil to the board. Is connected to a power source not regulated that goes from 5V to 12 V.

- **PWM**. It features 6 digital pins labeled as 11, 10, 9, 6, 5, and 3. These pins are specifically placed on the board to offer pulse width modulation (PWM). This procedure is often used to create results analog with digital resources.

- **RESET**. The plate Pro Mini comes with a pin reset which is very useful when the plate hangs while you are still running your program. You can comfortably reset the board by making the pin to appear low.

- **Head of programming.** The head of the six pins of the FTDI is connected in these pins and is used to program the board.

- **SPI**. Represents the Interface Peripheral Serial that is primarily used for the transmission of data between the microcontroller and other peripherals such as sensors and records. For this purpose, you can use four pins which include 13(SCK), 12 (MISO), 11 (MOSI), and 10(SS),.

- **Analog pins.** A0-A7 are the descriptions used to label the 8 different analog pins that are present on the arduino board. With a full resolution of up to 10 bits, you can use these pins to input analog signals.

- **External switches.** There are two switches available. They are called T1 and T0 and they are referred to as external switches. Hardware interrupts is also another term used to label these switches.

- **I2C**. A4 and A5 are used to develop the I2C communication. A4 is referred to as a line of serial data (SDA) that contains the data, and A5 shows the clock line serial (SCL) which provides the synchronization of data between the devices.

Differences with other Arduino boards

The most common thing with many of the Arduino boards is that they can send the program from your computer to the board with the help of a USB port present on them. However, in the case of the Arduino Pro Mini, all USB circuits are removed to make them as little and compact as they possibly can. Using a cable converter from USB to serial, you can also program the board. The module serial USB FT232RL is very handy and preferable to program this board. You can connect a head FTDI six-pin to USB converter series that provides power to the USB.

If you've already worked on the Arduino One, then it is not necessary to buy a converter cable USB to serial because you can program Pro Mini using the plate One. Make sure that the version Pro Mini with which you are working comes with an up-regulation of 5V it works at 16MHz as the Arduino One. The programming board for the Pro Mini 3.3 V lacks compatibility with the Arduino One. This makes programming of the 3.3V version of the Pro Mini board to appear difficult.

Another huge distinction is the form factor and thus makes the board stand out. Like we've mentioned a couple of time, the Arduino Pro Mini comes in small and compact design. This also makes the board great for most projects. Howeve, there is a limitation to its compact and small size. It is not compatible with Arduino Shields unless you wire the board with Arduino Shield.

Where to use the Arduino Pro Mini

All the boards of Arduino are popular for their ease of understanding and application. It is also an open-source platform where you can get all the related data and the schema of the original modules. Depending on the need or requirements of your project, you can customize your system with this platform.

In the market today, you can find a large range of arduino boards. Each one comes packed with a lot of features and packages. One can choose the plate suitable depending on the need. There are few cases in which we choose the Arduino Pro Mini instead of another:

- **Case 1:** Where the system is permanently installed. In applications permanent, the plate needs only to be programmed once and that's it. In such cases, the features provided as the USB programmer, the I/O connectors, and other hardware support are useless. The PRO MINI is designed specifically for those systems. This motherboard only has the basic hardware that is enough for these applications.

- **Case 2:** For convenience. This card is one of the smaller Arduino. With its comfortable size, it can be used in mobile applications.

- **Case 3:** the basic hardware and the cost of the plate is considerably less.

- **Case 4:** With 32Kbytes of memory the Pro Mini can accommodate most of the application programs.

Applications

- Projects hobby.

- Systems of energy supply.

- Applications of IO.

- Display systems.

Chapter 3

Arduino programming fundamentals

In the tech world, there is always a sense of uncertainty and you must learn to live with it. There is a premise that you must apply when you are doing an Arduino task. You have to be willing to assume that even if you don't understand what you are doing, you must continue towards your goal.

You must be able to move forward, not get stuck in some concept. It doesn't matter if you don't understand it, the important thing is to move on.

You will see how over time, what you did not understand will be seen more clearly from another perspective.

The syntax of programming with Arduino

In this section, as in any other, I like to put analogies so that everyone can understand. It is a way of explaining something in ordinary words so that you can learn Arduino. Contrary to what it may seem, there is a close relationship between the grammar and punctuation of a language and the syntax of programming.

I like to say that if you can read and write, you can program. You are probably familiar with punctuation and grammar. You will know what a period, a comma, hyphens, accents, or semicolons are. In addition to all this, a language has verb forms, nouns, and all kinds of grammar.

The goal of punctuation and grammar is that people can communicate through the written word. We use the comma to pause, a period to end a sentence, and the bold to emphasize.

In programming languages, the syntax is required to communicate with the compiler. The same goes for a language, punctuation marks and grammar are essential to understand each other.

The compiler will read that code and translate it into machine code for the microcontroller to understand.

We can say that the compiler will be your high school language teacher. The difference is that this teacher is the most demanding you have ever met since he always requires a 10 in your programs :).

He won't let you pass a single one. It's not like writing an email to a friend where it doesn't matter if you put an extra point or don't put an accent. The compiler will always demand that you have everything perfect.

Don't worry if it costs you at first, it's normal. Little by little you will understand what the compiler is telling you and where the errors are. I assure you that if you show interest, it will become your second language after your mother tongue.

Now we are going to see the most important particularities within the syntax of programming with Arduino.

Comments in an Arduino program

The comments are nothing more than notes that the programmer leaves within the code. It helps you understand part of that code. An important notice. It is not about making a complete report on what something in the code does, it is about putting something descriptive to help understand it. It is a fundamental part when we are writing a program for Arduino. However, it is not compiled, that is, it is not translated into machine code. Every time the compiler encounters a comment, that line is skipped.

There are two ways to write a comment.

- Everything to the right of the double slash (//) is considered a comment and is grayed out. The Arduino IDE changes color to indicate that it is a comment.

- The other way to put a comment is by putting / * to open and * / to close. Everything between these two open and close marks is considered a comment. As such, its color changes to gray.

Semicolon

In the programming language C ++, the semicolon (;) is like a full stop. Basically what it is saying is that we have finished a sentence and from that moment, we start something new without relation to the previous at the syntax level.

Surely it may not mean anything to you if you have never programmed before. The compiler interprets that from that semicolon everything you write will be a new statement, with nothing to do with the previous one.

At this point, we can do a test to see how the compiler works. Open a new program in the Arduino IDE and copy the following code. You don't even need to connect the board to your computer.

Now click on the verify button, it is the first shortcut in the editor. What happened?

```
 2   int variable1 = 0

 3

 4   void setup() {
 5       // put your setup code here, to run once:
 6
 7   }
 8
 9   void loop() {
10       // put your main code here, to run repeatedly:
11
12   }
```

```
expected ',' or ';' before 'void'

sketch_may29a:4: error: expected ',' or ';' before 'void'

 void setup() {


exit status 1
expected ',' or ';' before 'void'
```

Let's analyze it to see how to solve it.

If you look at the message area, the error is very descriptive, " expected ',' or ';' before void "that is, we need to put a comma or a semicolon before the word void.

In the console, it gives us even more information. It indicates the file where the error has occurred (sketch_may29a you may have another name) and the line where 4 has occurred.

Finally in the editor, we get a red stripe indicating the line where the error is and where it is referenced in the console. Now change the code and put the semicolon at the end of line 2 and you will see how it compiles.

As you can see, you are not alone. The compiler is very demanding but it also helps us to correct errors.

Reserved words

Surely you have already copied and pasted some of those Arduino codes that you find on the Internet. In doing so, you may have noticed some words change the text color within the Arduino IDE.

They are words reserved by the C ++ language. Each of them has a specific function and is reserved by the programming language. This means that we cannot use those names to name a variable for example. There are more keywords than we have seen. Little by little you will become familiar with them as you need them. The important thing is that if it changes color in the development IDE, it is a reserved word.

The functions in an Arduino program

Functions are one of the essential parts of a program. Although it seems somewhat advanced for this Arduino guide.

A function is nothing more than a piece of code that is used frequently within a program. The objective is to facilitate the use of this piece of code and to make the program cleaner and more readable. Imagine that in a program you are developing you need to know the area of a circle. You use this operation in many places within the code. To calculate it you just have to multiply pi by the square of the radius.

Now suppose you have to use this line of code many times in your program. You may be interested in turning it into a function. Arduino incorporates a lot of functions. They are quickly located in code because they change color, like reserved words.

Now we are going to focus on syntax when calling a function. All function calls begin with the name of the function itself. It is then followed by an open parenthesis and if the function has parameters they would be included below, separated by commas.

Finally, we close the parentheses and do not forget the semicolon indicates that we have finished with the sentence. Also Parameters are information that we pass to the function to do its job. In the example we have seen to calculate the area of a circle, the function needs to know the radius. So we pass it on as a parameter.

If the function does not accept parameters, it is necessary to put the parentheses. And this is all you need to know, in principle, about functions.

Variables in an Arduino program

One of the most important concepts in programming is variables. I assure you that when you understand how they work, you have one of the keys to learning Arduino. A variable is nothing more than a programming tool that helps us store and retrieve information in our programs. And when we talk about information, what comes to mind? Well that is, memory :)

Using Arduino memory

The computers and the microcontroller in Arduino usually have a thing called memory. But what is memory? It is something that allows us to save information so that we can later retrieve and use it.

It is extremely useful. Imagine the example of the water level meter in a container. You want your project to be able to control the water level and that the LCD shows you the maximum value that the water has reached.

If we want to know what the maximum water level is, we will need somehow to save the information on the current level and the maximum level so far.

At this point, we have two questions, how to save this information in a program? And how do we remember where this information is stored?

We can liken memory to a bookshelf. We store things in every hole on that shelf. Each hole also has a number. Precisely, that number would be the variable.

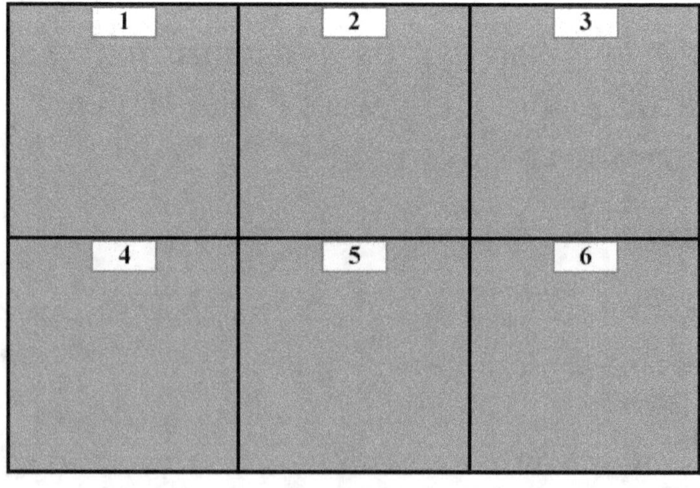

I have already told you that we have to store two data: the current water level and the maximum level that it has reached. So on this shelf; we are going to reserve two holes for those two data.

For example, we select slot 1 for the current level and 5 for the maximum level.

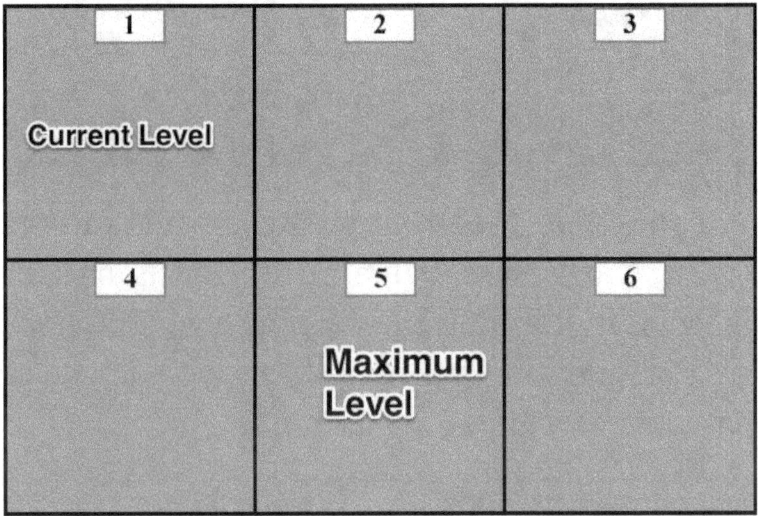

We start the program and it gives us a water level of 100 ml (milliliters). This would be our current level. At the maximum level, we have nothing stored and when comparing the maximum level (no ml) with the current level (100 ml), this level is higher.

Therefore this value will also be stored at the maximum level.

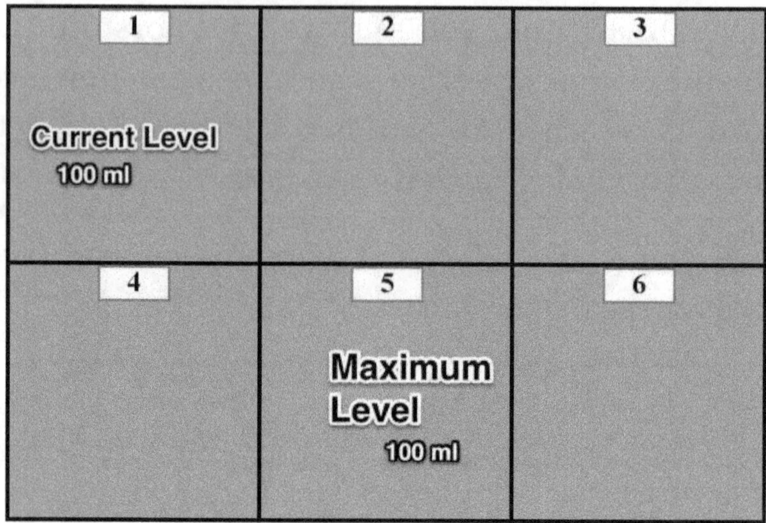

After some time, the program takes a measurement again. In this case, a value of 120 ml is obtained. This value is stored at the current level but since it is greater than the maximum level we had, we also update this variable.

What we have done is change the value from 100 ml to 120 ml. Here is the most powerful function of a variable: we can change the value of a variable and the name remains the same since the variable only contains the information.

We must stop for a moment and reflect on what we have just seen. What this comes to say is that what we are changing is the content of the shelf space but the name always remains.

The same is true for variables. You access the content of the variable and modify it but the name is always the same. That is, the name does not refer to the content that is inside, the name refers to the memory space that is reserved for storing information. If you know the name of the variable, you can put things inside that variable to later make use of that information. Remember, the name refers to the location, not the content inside

Type of data

Besides giving the variable a name, we have to do something else. We have to indicate the type of information to be stored. Imagine the cutlery drawer in your house. Normally when you remove the dishwasher and put the silverware in its drawer, each one goes in its place.

It doesn't make sense for you to stick your knives into the recesses of coffee spoons or forks with your knives.

The logical and normal thing is that you classify them. The knives with the knives in a recess adapted for them. The forks in another hole and the coffee spoons in their special hole for them.

If you do it another way, you know it won't work. When you go to look for a fork, you will not know where it is and it will probably take longer to pick it up. Where I want to go is that we must have a hole inside the drawer for each type of cutlery. The same happens with variables, we have to specify the type of data that we are going to store in each memory hole.

If you declare a variable that can only store whole numbers, positive and negative, you will not be able to store a number with decimals like 3.14. This would produce an error. Therefore, each variable can only store a specified data type when that variable is declared.

Declaration of variables

When I talk about declaring, what we are doing is creating a variable. To declare a variable you will need two things: a name and a data type.

In C ++ you always put the data type first and then the name.

```
sketch_feb21a §
int switchState;

void setup() {
  // put your setup code here, to run once:

}

void loop() {
  // put your main code here, to run repeatedly:

}
```

In the previous code, I have declared a variable of type integer (int) whose name is switchstate. The int data type allows you to store whole numbers between -32,768 and 32,767. This is called the range.

You do not need to memorize this information :). The word int is reserved by the C ++ language and as we have seen before it changes color in the Arduino IDE.

By the way, don't forget the semicolon at the end to indicate that you are done with declaring the variable.

To name a variable you have to follow certain rules. I've been programming for many years (more than 15 years) and I've seen everything.

Here are some basic rules that you have to follow to name a variable:

- They cannot have special characters ($,%, &, /, », ñ, etc...).

- Numbers can be used as long as they are not placed at the beginning.

- Cannot use reserved words as a variable name.

Variable initialization

When I speak of initializing a variable, it is nothing more than assigning a value. All this we have to do once we know the type of data of course. We can even do it along the same lines.

There are two ways to initialize a variable, In separate sentences or in the same line or sentence.

With this, we introduce something new. So far what we have seen is that we declare a variable with the data type and give it a name. Now we introduce the equality operator, the equal sign (=).

A declaration of a variable with initialization to a value would look like this.

    switchstate = 25;

But we can also use the assignment operator to assign a value. As you can see, it is very easy to store information in memory. The first thing is to know what type of data we are going to save. Then we give the variable a name and assign a value.

With these steps, we would already have the information stored in the Arduino memory.

Data types in Arduino

Throughout this guide, we are looking at different concepts. Let's say we are joining ideas one after another. now let's dive into C ++ data types.

To learn Arduino we have to know the language in which it is programmed. We are going to see the 7 most important data types in C ++, but they are not the only ones.

Integer (int)

It is the most common type of data. In C ++, the int keyword is used to declare an integer data type. It can store a 16-bit number that is, 2 bytes (16 ones in binary 1111111111111111). In total there are 65,535 numbers but since it stores positive and negative numbers the range of int is from -32,768 to 32,767. Something to be clear about variables is that if you declare an int variable, you are occupying 2 bytes of memory. It doesn't matter if you store 1 or 30,000.

Another particularity that we will see throughout the data types is what happens if we go out of range, that is if I try to store the number 32,768?

Contrary to what it may seem, it does not produce an error and this is a real problem. What happens is that it turns around, that is, if we have a variable of type int with a value of 32,767 (the maximum positive number) and we add one to it, the next value will be -32,768.

If we add two the value will be -32,767 and so on.

The same is true below. If we have a variable with the value -32,768 and we subtract one, the next number will be the highest positive, 32,767.

I know it seems incredible but it is. When we declare a variable you must be aware of the value you are going to store.

Surely you do operations, addition, and subtraction, and you must be able to intuit the limits to declare a variable of one type or another.

Unsigned int

What if we know that we are only going to work with positive integers? In this case, we can use unsigned integers. The C ++ keyword for this data type is unsigned int. It uses the same memory as an invariable, 2 bytes (16-bit). The only thing that only uses the integer part is that its range is from 0 to 65,535.

The operation is the same, if we add one to a variable whose value is 65,535, the result will be 0. If we subtract one from a variable that is worth 0, the result is 65,535.

Long integer (long)

If you still need to store larger numbers, you can also use the long integer. In C ++ the reserved word for this data type is long. You can store 4-byte (32-bit) numbers. The range of values is from -

2,147,483,648 to 2,147,483,647 . You can store positive and negative numbers. However, you can use the unsigned long keyword to store only positive numbers. This gives these types of variables the ability to store very large positive numbers only. Be very careful with memory as it is limited. We must do a preliminary study to know what values we are going to store.

Decimal (float)

If you want to store a decimal number with the data types we've seen so far, you couldn't. For the numbers with decimals, we have the floating-point numbers. In C ++ the keyword float is used. To separate the decimal part from the integer part, use the "." It may seem strange since we use the ",". This is due to the Anglo-Saxon influence on programming languages. With this type of data, we can store really large values. The range is 3.4028235 x 10 38 to -3.4028235 x 10 38 . Of course, it only uses 4 bytes of memory (32-bit).

How is it possible that with a float we can store larger numbers occupying the same bytes, 4, than the long variable for example? It is due to how this type of data is handled internally. To store it use scientific notation. This is beneficial on the one hand, as it uses memory in a very efficient way. On the other hand, it is harmful since it applies a rounding without knowing very well how it does it.

In normal conditions, a float is good up to 6 or 7 digits including the decimal part and the integer part.

Byte (byte)

It is the smallest data type we have seen so far. In C ++ the reserved word byte is used and as its name indicates, it only occupies 1 byte of memory. The range of values that you can store is 0 to 255 unsigned. Values can be assigned in different ways. The most typical is a value in the decimal system but we can also assign a value in binary. We do the latter using the B formatter. This indicates that after this letter comes a number in the binary system.

Boolean (boolean)

It is the type of data that occupies the least memory, 1-bit. The reserved word in C ++ is boolean and can take two values or 1 or 0. I have put several examples to explain one thing. First of all, 1 equals true and 0 equals false. On the other hand, this type of data is the only one that does not comply with the rule when it goes out of its range. As we have seen before, when we go from the top or the bottom we would start over. With a Boolean type variable what happens is that whenever it is 0 it will be false and otherwise it will always be true.

Character (char)

The last type of data that we are going to look at in this Arduino guide is the character. It is represented by the char keyword in C ++ and uses 1 byte of memory (8-bit). It allows us to specially store a letter. We are storing a letter as if it were a number using ASCII. Thanks to a table where the correspondence between letter and number comes, we can know what character it is.

In an example, we see how we can assign a value to the variable of type char in two ways. With the letter in single quotes ('') or with the ASCII code.

And with this, we finish the basic data types in Arduino. They are not the only ones but they are the most used and the ones that you will work with the most, especially when learning Arduino.

Program Arduino with functions

I have already told you that functions are one of the most important parts of an Arduino program.

Although it seems somewhat advanced, it is what you are going to face first in any Arduino course.

A program starts with two empty functions, setup () and loop (). The sooner you face them, the sooner you will learn Arduino.

The functions have two main objectives:

- Keep the code clean and easy to read.

- Have quick and easy access to chunks of code.

Call to a function

Any programming language has huge amounts of functions (including C and C ++ on which Arduino is built). For anything that is repeated, there is a function. That is why it is essential now that you are in the phase of learning Arduino, to understand how to make a call to a function. It's very simple, you just have to put the name of the function and open and close parenthesis. Sometimes the parameters are inside the parentheses. At the end of the whole semicolon.

The digitalWrite function allows us to output 5V or 0V through a pin. To do this we have to pass the state information (HIGH would be 5V and LOW would be 0V) in addition to the pin we want to work on. This is a call to a typical Arduino function. You should get used to this term as soon as possible since you will use it on many occasions.

The parameters of a function

Some functions require parameters as we have seen in the previous example. For example, the digitalWrite function needs to know two things: the PIN and the state (HIGH or LOW).

It's not complicated, is it? Now surely you are wondering how do I know what parameters a function admits. In this case, I have told you but in other cases, it is not so simple. The most typical thing is to go to the language reference. It is as if it were a guide or user manual. Almost all languages have their reference. In Arduino, we can find it on the official website.

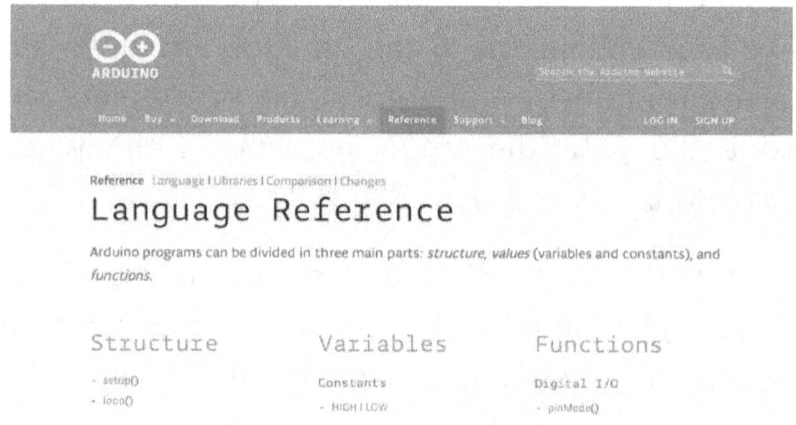

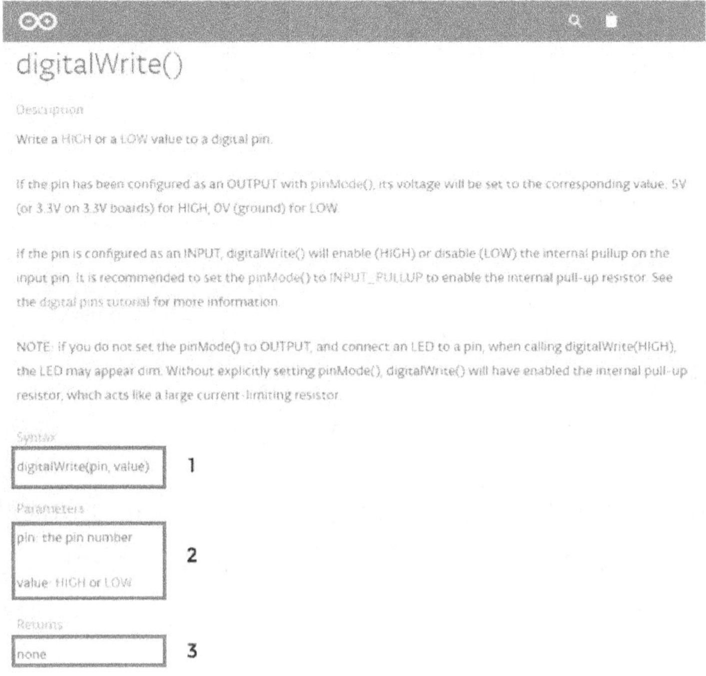

In the digitalWrite reference we find everything you need to know how to use this function:

- Syntax (syntax): how it is written and how many parameters it supports.

- Parameters: explains each parameter and what it means.

- Returns: If the function returns a value, here it will tell you the data type. The digitalWrite function does not return any value, so it puts none (nothing).

It is a matter of time before you learn the syntax and parameters used by the most common functions in Arduino.

Different uses of functions

So far what we have seen is a function, digitalWrite. It allows you to modify the state of a pin. If you put HIGH, it takes out a voltage of 5V through that pin. If you put LOW, it brings out a voltage of 0V.

What this function is doing is something physical. You are activating or deactivating some electronic mechanisms to supply 5V or 0V.

But there are other types of functions that give us information about a certain pin. For example, the digitalRead function. This function informs what state a pin is in.

To store the information of the state of the pin we use a variable of the type integer (int). This function can only return either HIGH or LOW.

The digitalRead function supports one parameter. In the example, we pass pin 5 as an argument. What this function does is check the state of the pin and return it to the variable.

We can see all this again in the language reference.

digitalRead()

Description

Reads the value from a specified digital pin, either HIGH or LOW.

Syntax

digitalRead(pin)

Parameters

pin: the number of the digital pin you want to read (*int*)

Returns

HIGH or LOW

Where we check that this function can only return HIGH or LOW.

Ok, so far we've seen functions that take parameters but some functions don't have any parameters. What this function does is calculate the milliseconds that have elapsed since the Arduino board started executing the program. It doesn't need any parameters. By itself, it knows what to do. Returns the number of milliseconds that have elapsed in a long unsigned integer as we can see in the language reference.

millis()

Description

Returns the number of milliseconds since the Arduino board began running the current program. This number will overflow (go back to zero), after approximately 50 days.

Parameters

None

Returns

Number of milliseconds since the program started (*unsigned long*)

But we can also create our functions. Imagine that you create one that allows us to turn on a light in the living room.

In this case, we do not pass any parameters and it does not return anything. The important thing in this section is that the functions have something more behind than the simple call. We can see it as if it were an iceberg. We only see a part of the function but below there is much more information. However, at first, you just have to know how to make function calls. Later you will be able to program yours.

CHAPTER FOUR

Arduino IDE

What is Arduino IDE

The software that allows you to program your Arduino board is called IDE, which stands for Integrated Development Environment. Indeed, this application integrates the editing of programs, uploading to the Arduino board, and several libraries. The IDE exists for all three operating systems and this article shows you how to install it on your computer. As of the publication date of this guide, the most recent version of the IDE is version 1.8.13.

Arduino IDE is a text editor and compiler to program and transfer the content of the instructions to the Arduino board in your machine language. The programming language used is processing. The Arduino IDE Software consists of 3 main parts:

Button panel or navigation bar:

- Verify: It is in charge of verifying the syntax of our program.

- Upload: If the verification was successful, we can upload the code to our Arduino board.

- New: We simply open an empty document (except the main functions) to start a new program.

- Open: To open projects in other directories or paths.

- Save: Simply save the program in the directory that we specify (if it is the first time we save it).

- Serial monitor: Suppose we need to know at some point what happens inside our Arduino board, well, through the serial monitor we can send data that will be displayed on our monitor.

- **Programming editor**: It is the main part of the Arduino IDE, basically where the lines and lines of code are programmed in processing language.

- **Notifications:** Usually known as the console, it is the debugging part where it notifies the programmer about syntax errors, communication, etc.

How to download and install Arduino IDE

It is recommended to install the IDE before connecting an Arduino board to USB for the first time; thus the necessary drivers will already be installed. The software is written in Java. The installation of this tool depends on the operating system of your computer. We will therefore detail the different steps for each system.

Download Arduino IDE from the official site

The first thing to do is to go to the site www.arduino.cc or www.arduino.org which both offer the same version (1.8.13) of the software, in the "Software" tab for the first and "Download" for the second.

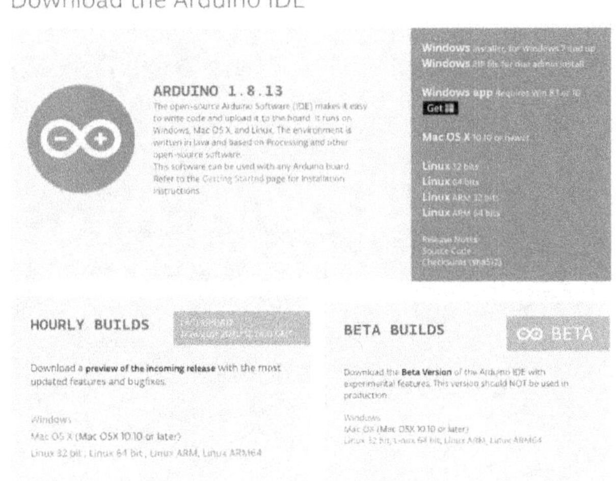

Figure 1 shows the part of the "Software" tab allowing you to choose the IDE according to its operating system.

Figure 1

Taken from the Arduino.cc site

If you use the Arduino.cc site, once you have made your choice of operating system (see what is explained below), you will come across a page that offers you to donate to help with the development of the software: "contribute and download". If you do not wish to donate, you will have to click on: "just download".

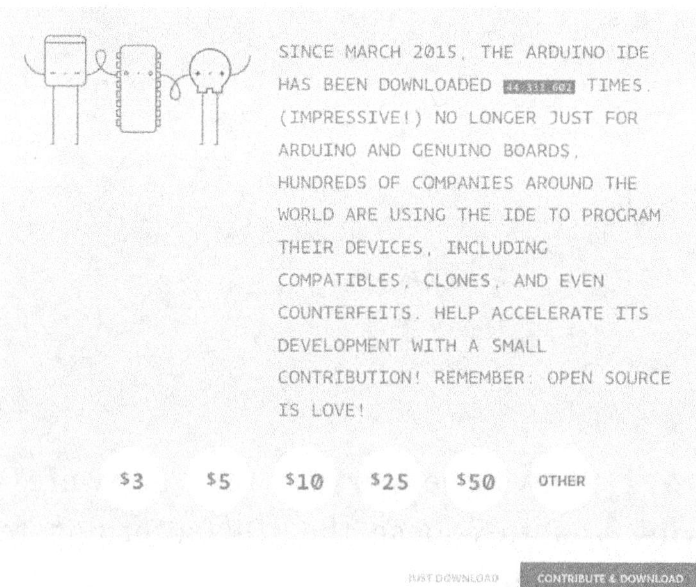

Windows

The IDE requires the presence of the Java Runtime Environment to function. To find out which version of Java is installed on your computer, you can visit this site: https://java.com/en/download/installed. jsp and optionally install or update Java. The software is suitable for Windows XP to Windows 10. If this is the first installation of the IDE on your computer, we advise you to choose in the "Software" tab the option "Windows Installer" which is an installable version; you just have to follow the various steps and the drivers for the Arduino boards will also be installed. This requires having administrator rights on the computer you are using.

Another possibility is the "Windows ZIP" option for those who are not computer administrators. You will get the download of an "Arduino-1.8.13-windows.zip" file that you save wherever you want; then you just have to right-click "Extract all ..." to obtain an "Arduino-1.8.13" folder in which the arduino.exe executable program is located (not to be confused with arduino_debug). Create a shortcut to get the IDE icon on your desktop.

Please note, if the installation by Zip file makes it possible to coexist several versions on the same machine, this is no longer possible with the installer. If an older version is already present on your machine, you will have to uninstall it before upgrading to the new one! On the other hand, an installed version can very well coexist with one or more versions deployed by the Zip file. For this reason, the installation by the ZIP file is interesting if it is not the first installation and allows us to install a portable solution.

At the first launch, the IDE will create an 'Arduino' directory in your 'My Documents' (or 'Documents' folder on Windows 10) which will contain your future projects.

MacOS X

The file to download is a .zip archive. Once this archive is decompressed, which is done automatically if your browser settings have not been changed, the Arduino application icon is located in the download folder (the desktop most often if it is Safari).

You get the Arduino app which you can put anywhere you want. We recommend that you place it in the 'Applications' folder, and then add the application icon to the Dock, to access it as easily as possible.

At the first launch, the program will ask you to accept the fact that you downloaded this application from the web (and not from Apple which securely protects its applications). Click on the 'Open' button. You will not be asked the question again.

Then, the IDE will create an 'Arduino' directory in your 'Documents' folder which will contain your future projects as well as the libraries that you will add as and when you need it.

Warning! The Arduino 1.8.x application contains the Java 1.8 runtime. It does not use the Java that you may or may not have installed in your Mac previously. Sometimes it is desirable to keep the original Apple JRE 1.6 for compatibility with other Mac OS X applications. So don't change anything to Java in your Mac for the Arduino.

If you have used previous versions of Arduino and are having problems, we recommend that you delete the old preferences, which are in the folder:

/ Users //Library/Arduino15/preferences.txt And restart the Arduino application again. If you have an older Mac that only supports an older OS X system, you will find the Arduino version that is suitable for your Mac and its systems.

GNU / Linux

Usually, Arduino can be found in various Linux distributions but you won't have the latest version of the IDE. The best is therefore to download it from the Arduino site, choosing the one that corresponds to your version of Linux (32 or 64 bits). The installation steps can be a little different from one distribution to another. The download retrieves a compressed archive "Arduino-1.8.13-linux64.tar.xz" in the download directory "Downloads".

Right-click on this archive and then select "Extract here" (extract in this directory). We then obtain an "Arduino-1.8.13" folder which includes a "shell script" file called "install.sh" which will allow you to install the Arduino IDE. From there, your best bet is to switch to Terminal mode. Go to the "Downloads" directory by typing: cd downloads (+ Enter), then in the same way in the "Arduino-1.8.13" directory by typing: cd Arduino-1.8.13 (+ Enter).

Once in this directory (where the script is located), type: ./install.sh (+ Enter). After a very short time, you will see that the operation has been carried out (done!); you just have to type: ./arduino (+ Enter) to start the Arduino IDE.

At the first launch, the IDE will create a 'sketchbook' directory in your 'home' which will contain your future projects.

Driver installation in Windows

The installation of the drivers is done at the same time as the installation of the IDE and the original Arduino boards are generally well recognized as soon as they are connected via USB. However, Arduino board clones can use different electronic components which may require the installation of additional drivers. This is particularly the case for clones that use a CH340 interface to communicate via USB.

Check after installation

Launch the freshly installed IDE. In the File menu, find the Blink program given as an example (File -> Examples -> 01. Basic -> Blink). Connect your Arduino board by USB to your computer. If your board is an Uno board, you should read at the bottom right of the IDE "Arduino/Genuino Uno on COM4" (Uno and 4 are examples). Before uploading this program to your Uno card, check in the Tools that the type of card is indeed Uno and that the port is COM4, otherwise change the options. Upload the program and observe the LED flashing at a frequency of 2 Hz (one second on and one second off). If this is the case, the IDE is perfectly installed and you will be able to program your cards.

Arduino IDE Online (Arduino Create)

Arduino Create is an online tool that allows you to program any Arduino board using a simple web browser (Chrome, Explorer, Firefox, etc.), without the IDE having been installed on your computer. Arduino Create also includes a storage space for your sketches, tutorials, a place to see projects made by other users, tools to facilitate the realization of IoT projects (Internet of Things), etc.

For this section, we'll limit ourselves to signing up for the service and using the Arduino Web Editor.

Signing up for Arduino Create

I hope this is only temporary, but signing up for Arduino Create seems unnecessarily complicated to me: you need to request an email invitation to be sent to you (which should reach you within 24 hours).) before you can do anything. No question, therefore, to start programming within minutes of your registration: you must wait several hours.

You start by accessing the Arduino Create site.

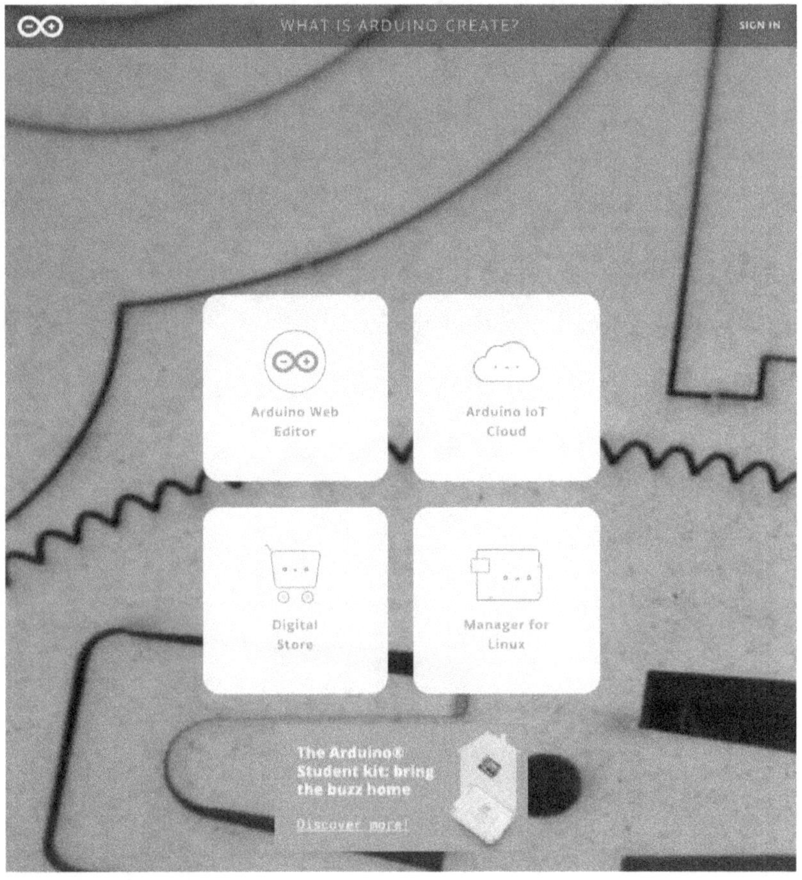

You must first have an account on the Arduino.cc site. If you already have one (for example, because you happen to participate in the discussion forum), you are fine. Otherwise, click on the "SIGN IN" button at the top right of the page.

You access the page that allows you to log in when you already have an account. To create a new account, you click on "Sign Up". Fill out the form choosing a username and password.

You will be presented with a very long text listing the conditions of use of the site. You must of course click on the "AGREE" button.

Installing the plugin

Hopefully (which isn't always the case), you should now be offered the option to download the plugin. The Arduino Web Editor allows you to program the Arduino without having installed the IDE, but you still need to install some files locally on your computer; otherwise you will be unable to upload anything to your board. So click on the "NEXT" button, then on the "DOWNLOAD PLUGIN" button

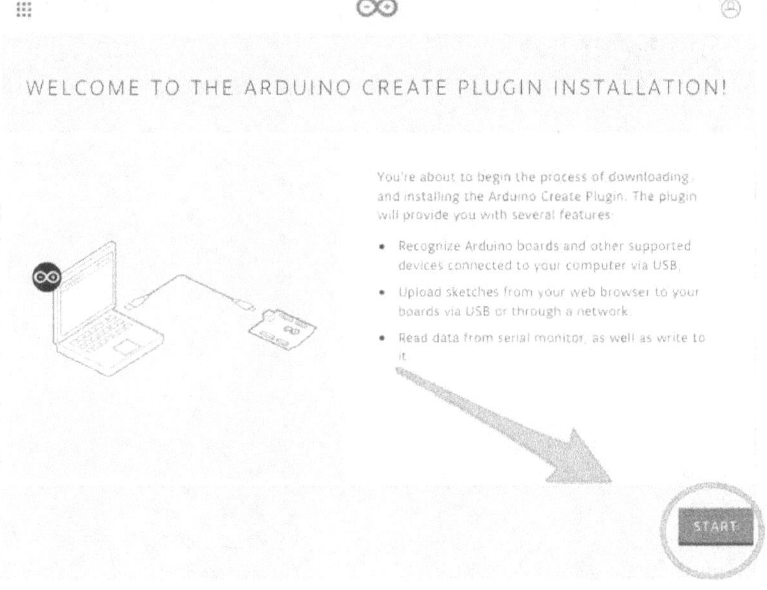

Find the installation file in your computers download directory (it is titled "ArduinoCreate Agent-1.1-windows-installer") and run it (a few

dialogs will appear to let you confirm the installation of various components).

After a while, your web browser should confirm that the plugging has indeed been detected. You must then restart the browser.

Using the Arduino Web Editor

So you can close Chrome, start it again, and then go to the Arduino Create home page. Click on the "Arduino Web Editor" button and you are finally in the online editor!

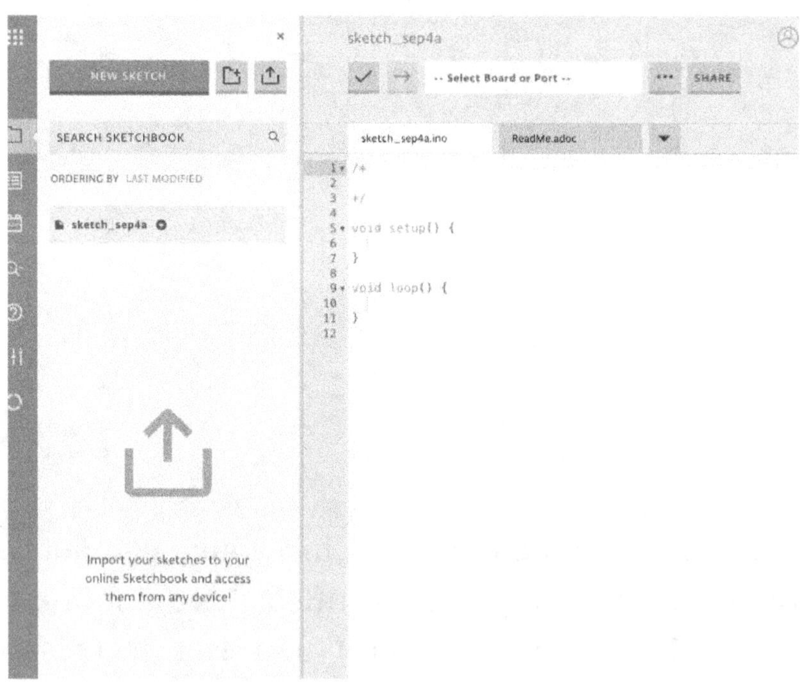

Not the same interface as the IDE, but you'll be able to find your way around anyway.

Plug a card into the USB port (a Chinese clone ... if it works with that, it will work with anything!). Click on "- Select Board or Port -" at the top of the screen, a list of available card models appear.

In the "Examples" section, select "Blink", click on the button which has the shape of a right arrow. If it works, the LED integrated with the card starts blinking.

Now for the serial monitor ("Serial Monitor" button). In this interface, the serial monitor is displayed in the same window as everything else. You can try writing a little sketch that displays increasing numbers in the serial monitor: it works great too.

About libraries, click on the "Libraries" button to see that the usual libraries are present. You will also see a button "ADD.ZIP LIBRARY" which should allow installing specialized libraries in the same way as is done in the conventional IDE.

CHAPER FIVE

Turning crazy or practical ideas into reality is something you can do with the help of the Arduino and the many compatible boards and accessories. And in a quite affordable way and with a DIY touch that ends up lifting your spirits.

The limit is your imagination and the desire to learn because the Internet takes care of the rest, as well as the Arduino accessories stores that flood the network,

Here are few Arduino Projects You Can Try :

Photo booth for your wedding or social event

Putting a photocall and allowing guests of a social event to take photos is fashionable. And there is a business around, as much as to cost about 600 euros to rent a booth to take photos.

With Arduino, a Dropbox account, and a webcam, you already have the basis to build your photo booth. And if you enlarge it so that it can print the photos that are taken instantly, success is assured.

Bubble making machine

A perfect example that, first and foremost, with Arduino is the imagination of power. Who would want a machine to make bubbles automatically? Surely who has a cat to piss off with this creation.

This does not mean that as a project with Arduino it is a little wonder and something to show off without a doubt. And once with the base, there are no limits to taking full advantage of a system where the protagonist is a robotic arm.

Your automated garden

One of the advantages of Arduino is the modularity that it allows us. Controller board and sensors that abound on the Internet and let your imagination fly. With temperature, light and humidity sensors we can configure an automated garden that is very attractive.

Back to the Future sneakers

It is not the most practical project that you are going to see in this selection, but it is the one that you might like to carry out the most. If you've ever dreamed of Back to the Future sneakers, Arduino lets you create them yourself by adding motors.

They are not very wearable but what the hell! They tie themselves and that is more than enough.

An indicator jacket for cyclists

Leah Buechley created a version of a special Arduino board to implement her idea: a jacket with lights to indicate the direction that we are going to take to cars or other cyclists thanks to the movement of our arms. Its plate, called LilyPad, is specially prepared to be sewn and integrated into garments without making them uncomfortable.

Shutter for high-speed photography

Taking the perfect photo at high speed can also be done with the help of Arduino. In this project, you will be able to build your self-timer to freeze the action at the precise moment, which will determine a system that will activate the camera at the time you have programmed it.

Home alarm

One of the most affordable and striking projects is the construction of a home alarm. With the help of motion detectors and many lights and sound, if we want, we can build our own, not very valid to protect the house but to be alert in case our pet decides to invade a protected area of the house.

Your affordable Ambilight

After managing to implement a basic Ambilight system a few years ago, the creator of this project has updated the system to get from the hand of Arduino a more faithful response to the home lighting system. Analyzing and detecting color changes in the source, mainly software that acts as a multimedia center on a computer, we can get a system that is quite faithful to the original Philips. And all for about $60.

A spy camera

It is not very discreet but it can strain. And whether you use it as a spy or not, just getting to work to create this coffee cup with a camera is already an adventure that you will surely enjoy. And quite a challenge to be able to fit the components in such a small space.

The spy camera uses an accelerometer module that detects when we turn a part of the coffee glass to activate the shot and voila, a photo without anyone (almost) noticing.

Intruder alarm in Minecraft

In Minecraft you cannot get lost or when you return to your favorite server, someone will have entered your territory. So that you find out if that happens there is a project that uses the Spark Core board and its WiFi connectivity.

With the ComputerCraft mod, we can generate a warning in the game when someone enters our territory and a "real" alarm is generated.

Build an Infrared Shooting Arcade

If you can finish it and make it work, this is one of those projects that all your visitors should see. It is a fairground attraction where you make even the gun with the help of LED lights. The rest of the elements allow you to unleash your imagination and show how far you can go. Oh, and of course, it's an Open Source project.

Notification of a new email

Although there are already commercial solutions that do it (the combination of IF and Philips HUE, for example), in this selection, you can try to make yourself a system that visually or audibly indicates the arrival of a new email to your account on the computer.

With the ubiquitous Arduino board, in this case the most basic one serves us, a USB connection with the computer and two small programs (one to check the incoming email and the other to activate the notification), you can create a new email notification system that can be made up of a simple LED, a speaker or whatever you want.

If this luminous cube falls short and you want something cooler for your desk, the next level is to have a robot that raises its hand every time you receive that long-awaited email.

Arduino Microcontroller Process Automation

Automation in modern society is a necessary measure because in the digital age, it is extremely important to exclude the human factor in various industries to standardize and improve the quality of products. There are some areas where people simply cannot control what robots are capable of producing, for example, nanomaterials and chips.

However, not only does automation help production, but it also helps the layman. For example, automating an Arduino brewery can make the production process much easier. Let's see how automating rectification on Arduino and other things can help, and consider some examples.

The main advantages of automated systems based on the Arduino microcontroller

Nobody forbids you to solder your board and program it yourself using low-level languages. However, automation on Arduino and finished microcontrollers will make the whole process much easier and save time. After all, it is much easier to buy an already finished product with a set of libraries and adapt it to your tasks. And the affordable automation of the Arduino Mega 2560 can be useful in many areas of life, from voice switches for a smart home to the electric devil with a motion sensor. The main advantages of Arduino automation are famous:

- **Low entry threshold.** It is not necessary to take an engineer training course, just watch some training videos and have basic programming skills.

- **A large number of libraries are already prepared.** The Arduino is used in the IEC by many robotics enthusiasts, to the point that the production of various electronic devices becomes their hobby. As a result, the user community is extremely active on the network, places a large number of blanks, and is ready to help you with any issues. The quality of the libraries, due to the low entry threshold, suffers, but no one forbids creating your own, just study the semantics of the C ++ language or use ready-made translators.

- **A large number of peripherals.** No matter if you need greenhouse automation for an Arduino or light sensor, you will find all the modules, right down to sound sensors and voice recognition. Yes, some of the cards cost a lot of money, but you can still find inexpensive analogs, for example, the wi-fi module from third-party manufacturer's esp8269, which costs 10 times less than the official one.

- **A lot of information.** Any problem you encountered was already with someone and you will probably find a solution for it in Google. There is also extensive documentation, which can be found.

However, don't think that Arduino is flawless. The board is famous for its poor performance. In particularly complex tasks and with a large amount of code, the response time can reach 1 second, which is not allowed for microcontrollers. The flash memory of most modules is no more than 1MB, which is not enough for creating neural networks or using media files. Of course, you can connect an auxiliary memory card, but it also increases the response time, takes extra resources for its power, and is done in a semi-homemade way.

However, simple automated systems, for example for beer brewing or greenhouses, do not require some of the resources that the council may issue. As a result, for most users, these shortcomings will seem unnecessary. If you decide to assemble your 3D printer or a more complex design, it is worth taking a closer look at the analogs. But the entry threshold for Arduino competitors will be much higher.

An example of Arduino microcontroller based process automation (Arduino Greenhouse Project)

The simplest example of process automation is the Arduino greenhouse. To create a system, it is necessary to clearly define the tasks that it must perform. Using the example of a greenhouse, this will be:

- Create a special climate.

- Timely on and off lighting.

- Quick watering of plants and keeping air humidity at the same level.

Based on these tasks, you can immediately notice what you need to buy on the main card:

- Temperature sensor. It will make sure that the air does not heat up or cool down, being within the limits prescribed by the program. In the event of a change in temperature, the card will include air conditioning or electronic batteries.

- Light sensor, Of course, you can limit yourself to a software solution and buy expensive lamps that imitate daylight. But if you want to make a full-fledged greenhouse, then it will be much more convenient to install an automatic ceiling, which will be controlled by Arduino.

- Humidity sensor. Here everything is the same as with the temperature, according to the prescribed scenario, the board will include sprayers and humidifiers, if necessary.

When you buy all the necessary modules, all that remains is to program them. Indeed, without code, they are only pieces of iron that are capable of nothing.

Arduino Microcontroller programming for process automation Example

As in the last paragraph, for programming, it is important to divide the task into separate sub items and execute them sequentially. Arduino programming is done through AT and AT + interface commands, using the prepared libraries. As a result, all scripts are written in a special C ++ environment and, before doing anything, spend time studying its semantics. Besides performing simple functions, the system is also capable of storing scripts in flash memory, which is what we need in this example. Keep in mind that the information for each sensor is provided in real-time and as variables, but you can limit the response time because there is no need to constantly spend resources and measure each parameter. Accordingly, adjust the time of each sensor on and off or adjust the response time for a certain period.

The Arduino setup and loop function

This is the other part that we are going to see in this Arduino course. Linking with what we have seen about the functions, I am going to tell you about the setup and loop functions.

They are the most important functions of a program and essential when learning Arduino.

Procedures of a restaurant

So that you understand the function setup and loop performed within an Arduino program, I am going to give you an analogy, surely you understand it better. Imagine you have a fast-food restaurant. You're the owner. You usually hire students from the area to work for a few hours.

More or less every month you have a new employee that you have to train. This leads you to have two procedures:

- Opening procedure.

- Customer service procedure.

The two procedures are in the restaurant so that employees know what to do at those times.

The opening procedure tells the worker what to do when the restaurant opens. For example, raise the lock, open the doors, put the open sign, remove the ingredients, etc ... This procedure is only done once a day when the restaurant opens. Once finished, it is put on the shelf for the next time the restaurant opens.

The second procedure indicates what an employee has to do when a new customer arrives at the restaurant. It is an enumeration of steps.

For example, ask how many they are, find a free table, take note of the drink, take note of the food, etc ... Therefore an employee will resort to the opening procedure once a day, and the customer service procedure each time a new customer comes to.

Right now you are thinking, what does a restaurant have to do with Arduino? Well, now you'll see, but an Arduino program has to do similar things to what is done in this restaurant.

The setup function

In an Arduino program, some things only have to be done once. Things like:

- Start serial communication with the computer through the USB port.

- Set the pins in input or output mode.

- Display a welcome message on an LCD screen.

- Rotate a servo motor to a default starting position.

- Start a counter to zero.

All these tasks have to be done only once. We can say that they are the initial conditions of the program. These types of tasks are those that go in the setup function.

Let's start looking at this feature in more detail. First of all, as with reserved words or data types, the setup (,) and function has a special color in the Arduino IDE.

This color indicates that the Arduino IDE identifies this function as a special type that is used to control the structure of the code. Like all functions, it has a parenthesis that opens and another that closes. However, this function does not need any arguments to operate.

To the left of the word, setup is the word void (meaning empty). This indicates that this function will not return any value or data.

Therefore, this function does not need any arguments and does not return a value but does something. Your job is to run all the code inside it. This is delimited with braces ({}).

The setup function is the first thing that runs when we launch the Arduino program. Once the setup function is executed, the program will go line by line executing each one of them.

The most important thing about this function is that it ONLY RUNS ONCE. This is very important; the code between those two braces will only be executed once.

It is like the employee who opens the restaurant, looks for the opening procedure, follows it strictly and once he has finished saves it and goes on to the next procedure.

The loop function

In an Arduino program, the next thing that runs after the setup function is the loop function. Its meaning is repetition and the function resembles the setup function.

It does not require any arguments but has the opening and closing parentheses. It does not return any value so it has the word void on the left.

Next, you have the opening and closing curly braces ({}). As with the setup function, when the loop function is executed, it runs line by line.

The big difference between the setup function and the loop is that when all the code in the loop function is executed, the program returns to the beginning and executes everything again. Get into an endless loop over and over again.

It is the same as in the restaurant with the customer service procedure. When the first customer arrives, all the tasks are executed to serve him. Then another client comes and the same tasks are executed again, like this indefinitely.

One of the most common questions is how fast does this function run? This is determined by each microcontroller. In the case of Arduino UNO, the microcontroller it uses is the ATmega328P and it has a speed of 20MHz (megahertz).

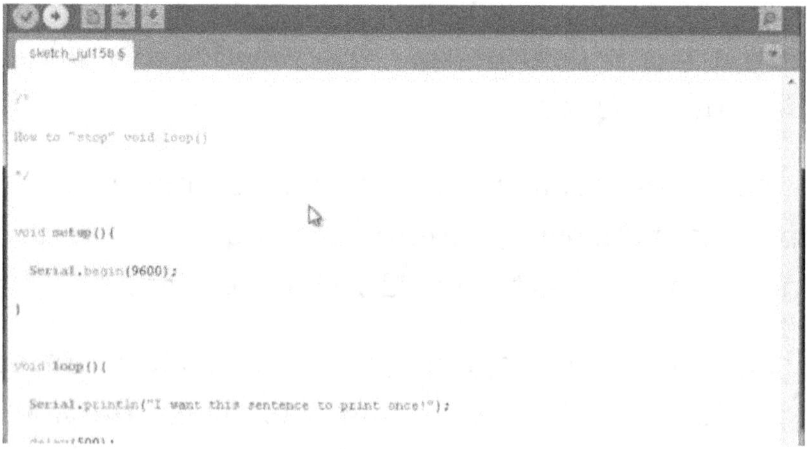

```
sketch_jul15a §

/*

How to "stop" void loop()

*/


void setup() {

  Serial.begin(9600);

}


void loop() {

  Serial.println("I want this sentence to print once!");

  delay(500);
```

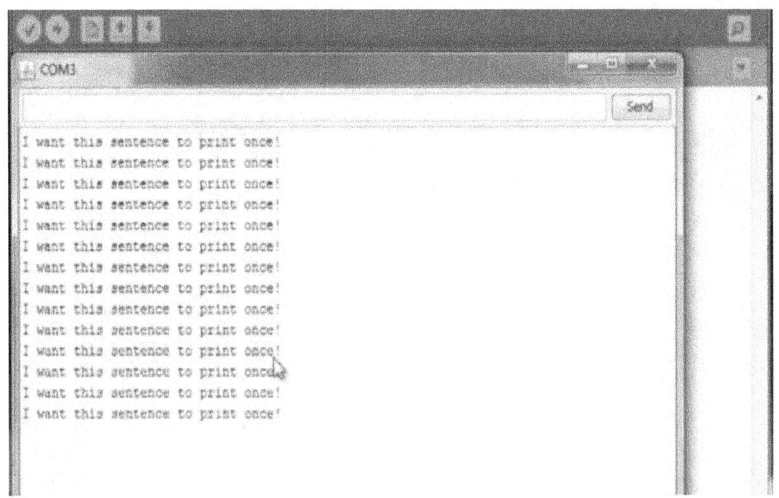

This equates to 20,000,000 instructions per second. Yes, you read that right, 20 million :). This is not to say that there are 20 million lines of code. I remind you that the code that you write is then translated into machine code.

Each line of code that you write can be translated into multiple instructions but even so, if we have 1,000 instructions it would take 0.00005 seconds. Can you imagine how fast it goes?

However, the restaurant analogy does not fit perfectly with the loop function. The employee will always wait for a new client to arrive to start the procedure. This does not happen in the loop function.

In the Arduino program, whenever the last line of code in the loop function is reached, it will go back to the beginning to rerun everything. It is not waiting for any external input to do it.

CHAPTER SIX

More Projects to Try

DIY Thermal Image
Introduction
How to take thermal photographs? This small inexpensive assembly compared to the price of a camera dedicated to thermal photography, is based on the technique of "light painting".

Step 1 - The assembly

Carry out the assembly as described.

Upload the attached program to your Arduino. To have visible results, i.e. blue colors for cold areas and red for hot areas, with good contrast, take measurements on your wall or object to have an order of magnitude of the temperatures to be detected.

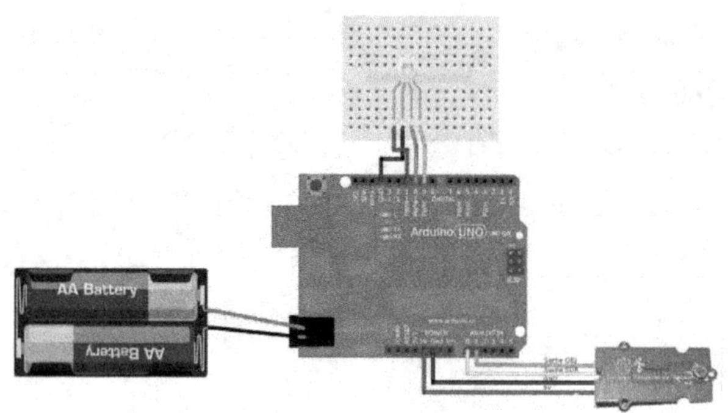

Step 2 - Setting the Min and Max temperatures in the code

and then modify the first two variables:

```
18 const  float  lowReading  =  0 ;  // low temperature in degrees (blue
)
19 const  float  highReading  =  45 ;  // high temperature in degrees (r
ed)
```

Once the assembly is done, tape the infrared temperature sensor well parallel to the RGB led.

Check with your hand in front of the sensor that the LED changes color, if this is not the case, control your assembly and/or modify the high and low-temperature parameters.

Step 3 - Adjusting the SLR Camera

Set your camera to Manual mode, then the aperture to F7.1 and the exposure time of 30 seconds ISO sensitivity 200

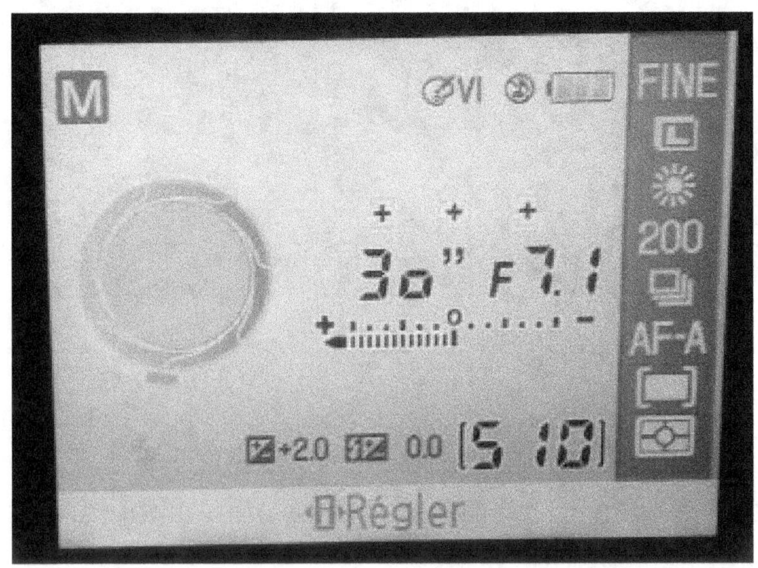

Step 4 - Shoot and scan

In a dark room, position your camera in front of the part of the wall to be "scanned". Trigger your camera, and sweep about 10cm across the entire surface of the wall.

Step 5 - Result

How does it work?

Observations: what do we see?

After 30 seconds of exposure, you should get a result like the one above

In a simple way, the program loaded in your Arduino senses the infrared temperature of an object and then transforms it into color. By placing a camera with relatively long exposure time and sweeping the wall with your lighting fixture, you perform light painting! But not just any light painting, you paint temperatures! The photo taken using your SLR is tinged with colors corresponding to the temperature of your object, red for hot temperatures, and blue for cold temperatures.

You have just taken a thermal photo!

Applications: in everyday life

According to Wikipedia, A thermal imager can be used in different situations. This list is therefore not exhaustive:

For firefighters

Search for victims during apartment fires and especially in large volumes such as underground car parks, factories, storage halls, forest fires, etc.

- hotbed search: the thermal camera can very quickly detect a hotbed or even a smoldering fire

- fire or residual hearth in an expansion joint following a cellar fire in an apartment building type residential bar,

- hot spot after a chimney or roof fire has been extinguished,

- electric fire: short circuit, false contact resulting in occasional overheating

- during the unloading of railcars or tanks, the level in the tank of certain chemicals can be observed using the thermal camera

- during an intervention for a night traffic accident in the countryside, to detect a possible body thrown off the road

- in clearing rescue, to locate a victim in a room accessible through a small opening

The models are generally not explosion-proof and therefore cannot be used in explosive atmospheres.

- For the army and the police services: for night operations;

- Recently, cinema operators in the United States have equipped their staff with thermal cameras to detect people filming screenings from the theater (screening).

- For the building:

  ✓ Detection of weak points in the insulation of a building.

  ✓ Checking the temperatures of pipes and heating installations, in particular for checking under floor heating.

  ✓ Checking of electrical cabinets by viewing overheating connections, or certain components.

- For airports:

  ✓ To check for people with a suspected fever.

- In the medical field (ex: Thermographs)

RGB Tape Control from Smartphone and Arduino

You can comfortably create your own home automation Arduino project where you can decide

The temperature and humidity. With the DHT-11 temperature sensor, you can also control the lighting thanks to the RGB LED strips and manage various devices wirelessly through the JY-MCU Bluetooth module from your mobile phone.

We are going to explain how with a few simple steps we will have our led control unit mounted to be able to use it and control the led of our home or business from our mobile or Tablet.

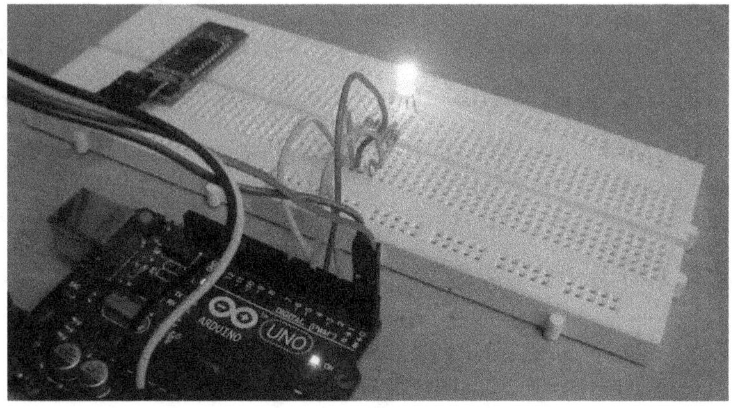

Step 1: Necessary Items

First of all, we are going to collect all the elements we need.

- Arduino (Uno, Mega or Nano)

- JY-MCU Bluetooth module (hc05 / hc06)

- ULN2003A Transistor Array

- 5050 RGB led strips Common Anode

- DHT-11 Sensor (Temperature / Humidity)

- 5v 4 Channel Relay Module

- 12v led power supply

- Software: Arduino

Next, we are going to assemble the electrical circuit as detailed in the diagram.

It is important to take into account that this scheme is designed to supply an intensity of 500mA for each RGB channel. (1 led strip of 1 meter per channel)

If you need to connect more LEDs, you will need a power amplifier that supplies enough intensity for your installation.

Once we have our circuit assembled, we are going to proceed to program the microcontroller.

Step 2: Program Arduino

To program our Arduino you must have the software installed and load the arduino code socket. Once loaded, you have to wait about 10 seconds since the Bluetooth module is programmed for its first use.

The end of the Bluetooth programming process has ended when channel 1 LEDs change from Red to Green.

Once we see the green LEDs we have finished with the configuration of our control unit and we will have it available for use.

Finally, we have to install an application which will be used to control the led on our Android device. We recommend Color LED Controller

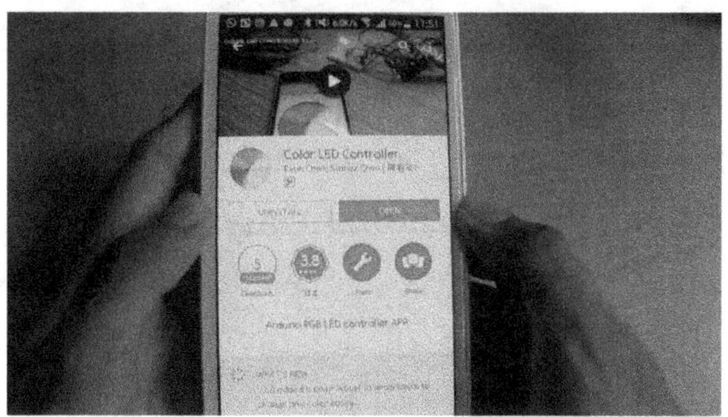

Step 3: Install the Color Led Controller App

We access Google Play and install it.

Once we open the application, we will be asked to connect with our Bluetooth device, we perform a scan and choose the application to connect. The default pin is 1234.

From the application options, we can modify the pin of the device to prevent other applications from connecting.

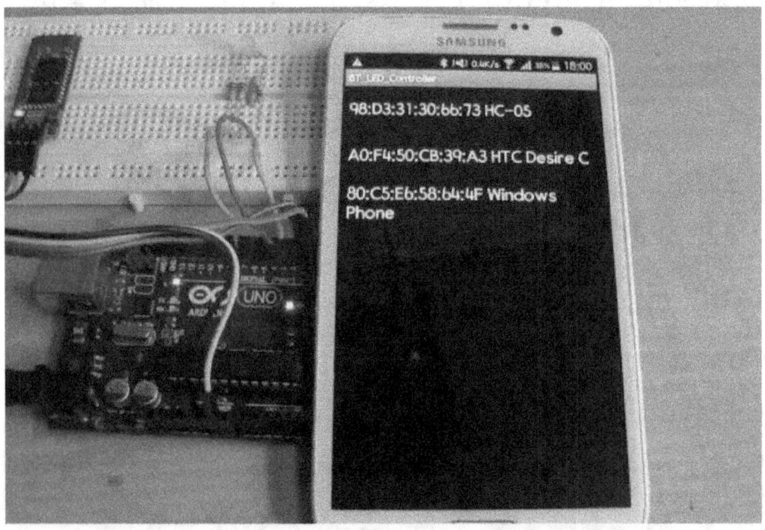

However, we will only be asked for the pin the first time to proceed with the pairing of our application. If the pairing has been successful, our application will change to the control screen.

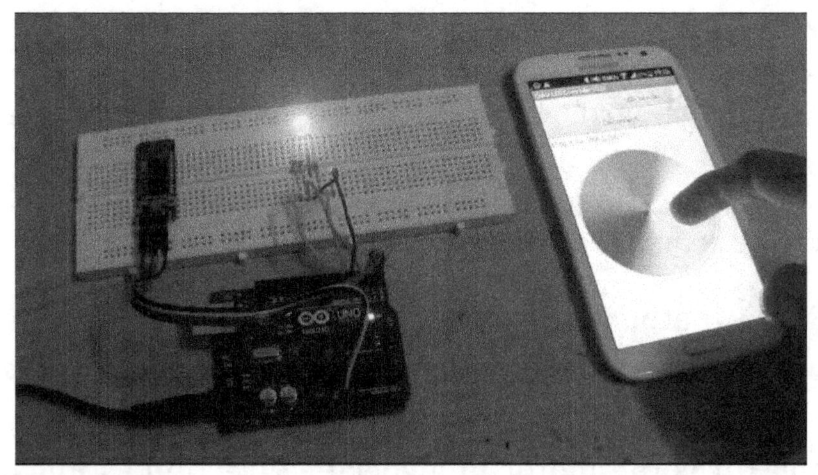

And this is it, we already have our Bluetooth control unit working.

Arduino Face Recognition and tracking system

Recently, as face recognition technology has developed a lot, many technologies apply face recognition technology. One of them is a technology called Face-Tracking that tracks faces and is widely used in CCTV. This is because a CCTV capable of Face-Tracking can efficiently show more ranges than a non-moving CCTV and can accurately capture a face on the screen. It would be more meaningful to implement Face-Tracking technology, which is applied in this way, and make it more meaningful if you make a robot that does not have a rigid camera and makes it look like a person.

1. Materials needed

Arduino board, Brad board, 2 servo motors, 2 brackets

What is a servo motor?

Servo motor is a motor that has three connection parts, one is GND (ground), the other is 5V, and the other is used as a wire to receive the input signal, and turns the motor in a specific direction as much as the input value. Here, it is used to rotate the camera by connecting to the camera and receiving the signal value from the Arduino.

What is a bracket?

A bracket is a supporting structure protruding from a wall or structure to hang equipment. Here, it's a part you need to hang the camera and steer the camera.

To track the face and turn the camera, you need to have the technology to track the face and rotate the camera through it. The overall mechanism is to recognize the face and rotate the camera so that the face is always in the center of the camera. Face tracking technology will be implemented through software and camera rotation will be implemented through hardware.

2. Breadboard & Schematic

Connection between Arduino and servo motor

To track and follow the face, the Arduino must receive the value to rotate the motor from the computer and send it to the motor. The picture shown below is a circuit diagram of two servo motors and Arduino using a breadboard. The red line is connected to 5V, the black line is connected to GND (ground), and the yellow line is connected to the input pin. The yellow line is the motor. This is the path through which input values that determine how much to rotate are sent.

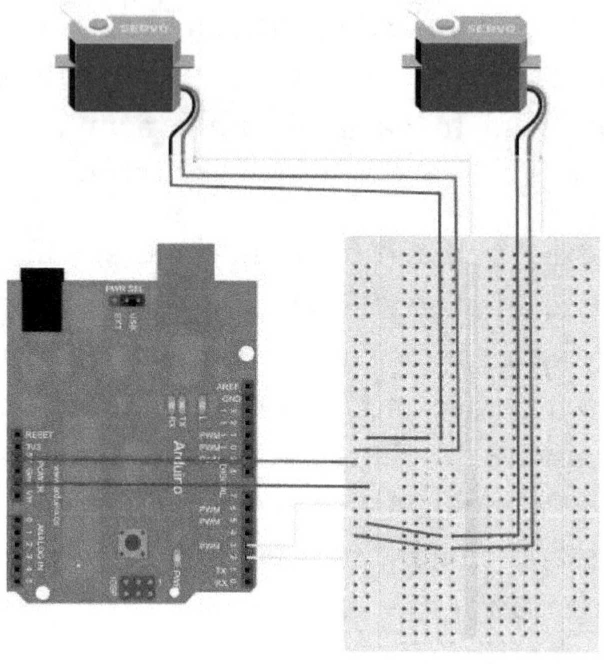

To follow the connection face of the servo motor and webcam, you need to attach a bracket to the servo motor and make a shape that attaches the camera to the bracket. At this time, since the face moves on the screen of the camera, you need to control both the x-axis and y-axis directions. So, two servo motors are used, one for the x-axis and one for the y-axis. The bracket and servo motor were connected with screws, and the bracket and camera were connected with a wire. <Connection part between bracket and the servomotor><Connection part between bracket and camera>

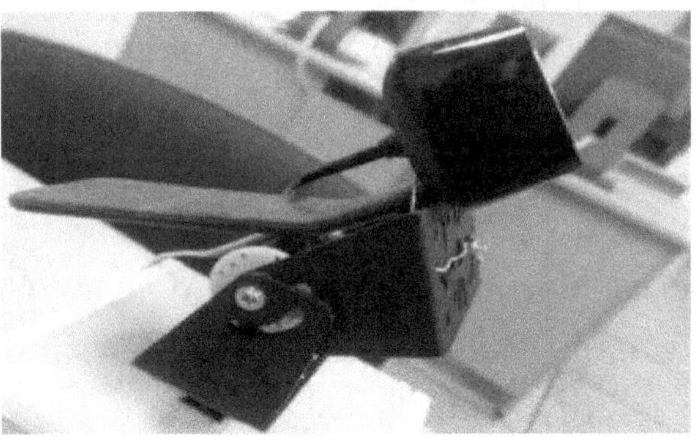

3. Code

The role of the software is to recognize the face from the image value input through the camera, calculate the value of the motor that the camera should rotate through the coordinates of the face, and transmit it to the motor. It uses three programs:

Processing, Arduino, and Open CV. Open CV is a program already known as a program that processes images. The processing uses this program to find a face in the image, calculate the coordinates, and send the motor value to be changed so that the face is centered on the Arduino using serial communication. Arduino transmits the motor value received from the processing to the connected servo motor.

Open CV installation and setup

First, you need to download the Open CV 1.0 version. http://opencv.org Download version 1.0, not the latest version here. If you look at the attached file, there will be 2 XML document files. These two XML documents are the primary sources for face recognition. In other words, Open CV accepts the information of the video and judges whether it corresponds to the value in this XML document and determines whether it is a face or not. There are many numbers written in the XML document. If you have installed Open CV, you need to set environment variables and put the Open CV file in the processing file of my computer.

Arduino software

```
#include <servo.h>
char tiltChannel=1, panChannel=0;

Servo servoTilt, servoPan;

char serialChar=0;
int inValue = 0;
void setup(){
  servoTilt.attach(3); //motor initialization to the Central view so that the
  servoPan.attach(2); //that part.
  servoTilt.write(125);
  servoPan.write(65);

  Serial.begin(57600);
}

void loop(){
  while(Serial.available() <=0); //Channel value and read the motor values to a variable.
  serialChar = Serial.read();
  if(serialChar == tiltChannel){
    while(Serial.available()<=0);
  inValue = Serial.read();
  servoTilt.write(inValue);

  }
    else if(serialChar == panChannel){ //Channel value according to the motor to steer for that part.
      while(Serial.available()<=0);
  inValue = Serial.read();
  servoPan.write(inValue);
    }
}
```

4. Finished picture

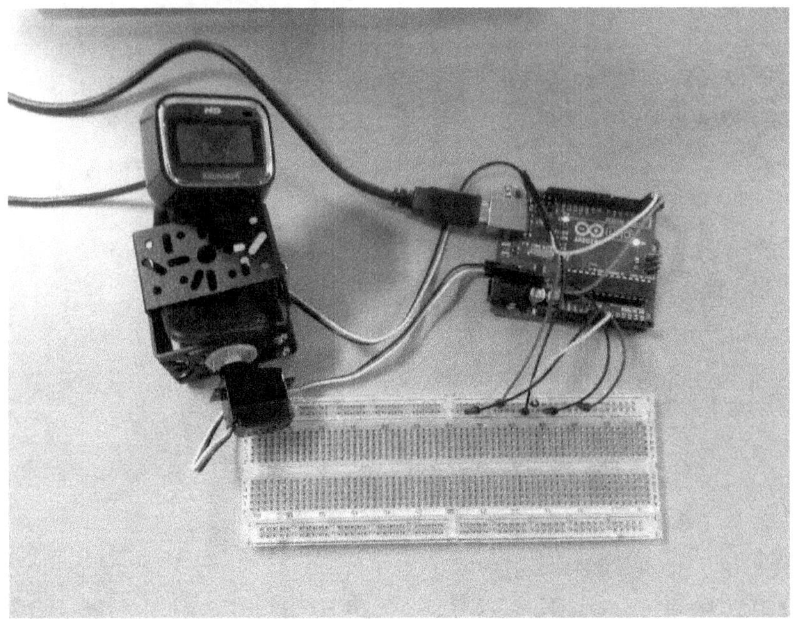

5. Conclusion

In the implemented hardware, by connecting the head of the robot to the webcam part and the body part to the servo motor part below, we created a robot whose body remains still and only the head turns toward a person. The webcam was inserted into the robot's mouth and fixed with scotch tape, and the motor below was inserted by drilling a hole suitable for the size in the body. The lower part of the torso has a wide plate for the center of gravity. Also, we used an algorithm that continuously rotates a certain value so that the face is in the center, but it did not appear to follow the face quickly. However, if too many rotations were given, the face was shaken back and forth when the face was still. So, I used P control in proportion to the error of changing the motor value more as the coordinates of the face and the center of the screen are different, and the video above is a video using P control. I felt good because I used P control in my project from PID control that I learned in school information class, and it was interesting and interesting because it was a work with cameras.

Create Radio with arduino

Surely you have in your storage room some RC radio of a Chinese drone that has stopped working, or simply a relative has given it to you because as you are makers you can surely take advantage of it. Well, you are in luck, in this section we are going to see how to modify and adapt an RC radio, take the guts out of it and leave our Arduino with the nRF24L01.

Making an RC radio ourselves with Arduino and nRF24L01 offers us several advantages and disadvantages that we will see throughout the entry. However, I can tell you that to control small home robots it is more than enough. Also, in the course of the project, we will learn to handle PWM signals, the Arduino ADC, and many other things!

How does a radio control system work?

Generally, all radio control systems work in a very similar way. All are based on emitting a radio frequency signal that contains the information of the signals to be received. The old radios emitted in VHF (around 35Mhz) and the only way to make sure that the signals did not collide in the air (and thus avoid interference) was to use quartz crystals. These were placed on the radio to analogically configure the exact broadcast frequency. With this technology, if two stations were broadcasting on the same frequency, the receiver would not know which signal to pay attention to and the dreaded interference would occur.

At present, signals in the 2.4Ghz range are mainly used, the same signals that your home Wi-Fi uses. Also, luckily, there are very cheap modules that allow us to send and receive signals on this frequency. The advantage of using this frequency range is that the bandwidths are higher, and more data can be sent per second.

The nRF24L01 module

The module I was talking about, which is capable of transmitting data in 2.4Ghz at very affordable prices (€ 1.5 for two) is the nRF24l01. Throughout this post, we are going to use two modules to communicate between the RC radio and the receiver. There are two variants of this module, the first, which works with an antenna placed on the PCB; and another with more power for more distant links.

Also, being well known, it has libraries available for the Arduino IDE, so the first tests and prototypes can be done very easily.

The RC radio with Arduino

The system that we are going to use is very simple, we connect the nRF24L01 to the Arduino, and on the other hand, we connect the analog and digital inputs that are connected to our buttons and control knobs. We read these signals and send them through the radio link.

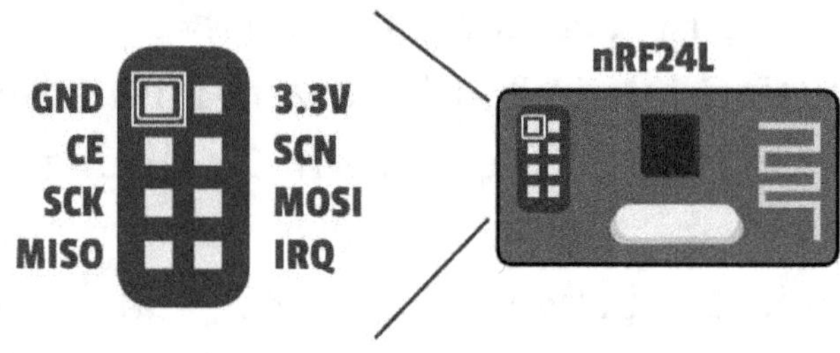

How to configure the nRF24l01 as an emitter

In principle, we are going to use the Radio to broadcast alone that is, we are not going to configure the receiver part to make things easier and to be able to do the project more quickly.

By using the library that we can download for Arduino, our entire configuration can be sumarized

```cpp
1.    #include <nRF24L01.h>
2.    #include <printf.h>
3.    #include <RF24.h>
4.    #include <RF24_config.h>
5.
6.    // RF24
7.    #define CE_PIN 9
8.    #define CSN_PIN 10
9.
10.   RF24 radio(CE_PIN, CSN_PIN);
11.
12.   const uint64_t pipe_tx = 0xAABBCCDD01LL;
13.
14.   void setup() {
15.
16.     radio.begin();
17.     radio.openWritingPipe(pipe_tx);
18.
19.      if (radio.isChipConnected())
20.     Serial.println("Connected!");
21.
22.     radio.setAutoAck(false);
23.     radio.setDataRate(RF24_1MBPS);
24.     radio.setPALevel(RF24_PA_MAX);
25.     radio.setCRCLength(RF24_CRC_8);
26.
27.      // Stop receive to only send data.
28.     radio.stopListening();
29.   }
30.
31.   void loop() {
32.      /*..*/
33.   }
```

We configure the shipping address, this address has to be shared by the transmitter and receiver. That is if one writes to the address

"0xAABBCCDD01" the other has to listen in the

"0xAABBCCDD01". Also, we deactivate the ACK, set the speed to 1Mbps, set the power to maximum, and activate the CRC.

Note: If you are using the USB to power this entire system, be careful because if it does not work for you it may be due to a power problem. Try the same using well-regulated 5V for the Arduino.

Sending structures to with the nRF24l01

Generally, all the tutorials I see use the nRF24l01 to send themselves very basic data or messages. It's fine to learn, but in practice, you are rarely going to send yourself strings (they are the devil in on-board systems), but you are going to try to send raw data, generally ordered. And the best way to send the data in an orderly way is by using data structures.

The way to send structs through the nRF24 is very simple. The first thing we need to do is define our structure, which will define how the data will behave within it. In our simplified case, as we are going to use only 4 ADC channels, the structure would be:

```
1.      struct msg_{
2.         uint16_t adc1;
3.         uint16_t adc2;
4.         uint16_t adc3;
5.         uint16_t adc4;
6.      } message;
```

That is, we have defined a message variable that contains four numbers that we have called adc1, adc2. We just have to fill in these fields, with the values of each Arduino Analog-Digital converter, and send the entire structure.

At the reception, we will have the same structure, and by simply receiving it, each field will have the value it had in the shipment. And how do we do this delivery together with the reading of our ADCs? Well, it is as simple as what you see in the following piece of code:

```
1.    #include <nRF24L01.h>
2.    #include <printf.h>
3.    #include <RF24.h>
4.    #include <RF24_config.h>
5.
6.    // RF24
7.    #define CE_PIN 9
8.    #define CSN_PIN 10
9.
10.
11.   // ADCs
12.   #define ADC_1 15
13.   #define ADC_2 16
14.   #define ADC_3 17
15.   #define ADC_4 18
16.
17.
18.   RF24 radio(CE_PIN, CSN_PIN);
19.
20.   const uint64_t pipe_tx = 0xAABBCCDD01LL;
21.   void setup() {
22.     pinMode(ADC_1, INPUT);
23.     pinMode(ADC_2, INPUT);
24.     pinMode(ADC_3, INPUT);
25.     pinMode(ADC_4, INPUT);
```

```
30.      if (radio.isChipConnected())
31.      Serial.println("Connected!");
32.
33.      radio.setAutoAck(false);
34.      radio.setDataRate(RF24_1MBPS);
35.      radio.setPALevel(RF24_PA_MIN);
36.      radio.setCRCLength(RF24_CRC_8);
37.      radio.stopListening();
38.    }
39.
40.    struct msg_{
41.      uint16_t adc1;
42.      uint16_t adc2;
43.      uint16_t adc3;
44.      uint16_t adc4;
45.    } message;
46.
47.    void loop() {
48.
49.      message.adc1 = analogRead(ADC_1);
50.      message.adc2 = analogRead(ADC_2);
51.      message.adc3 = analogRead(ADC_3);
52.      message.adc4 = analogRead(ADC_4);
53.
54.      bool ok= radio.writeFast(&message,sizeof(message));
55.      radio.txStandBy();
56.
57.      delay(50);
58.    }
```

And with this, everything would already be sending the structure with the four ADC values through our nRF24. However, not everything is beautiful. As in any project, everything has its advantages and disadvantages.

Advantages and disadvantages

The main advantages are the high possible customization of the command, as well as the data to be transmitted; and the possibility of implem-enting return messages to have some telemetry. However, on the other hand, the range is more limited than radios that are a little more profess-ional, and the reliability of the system depends a bit on the programming. Soif your software crashes; the system will stop receiving data and stop acting.

Create LED Cubes with Arduino

Today the process of making a 3x3x3 LED cube with Arduino is shown, highlighting the electronics and programming. I have watched some videos where you have 8x8x8 cubes. In this case, to make it accessible to people who also wish to do so, it has only been done with 9 LEDs on three levels. In this project, you will see some concepts of basic electronics and Arduino.

Let's do it!

Level: Basic

Cost: money> 20 USD.

Materials:

  5 27 LEDs

  6 3 NPN transistors (2N3904 or also 2N2222)

7   3 10 kΩ resistors

8   9 220 Ω resistors

9   1 Breadboard

10 1 Arduino UNO

11 Connection cables (4 ft will suffice) or Arduino Jumper Wires

Step 1: creating the circuit

When buying materials, look to buy LEDs that have sufficiently long pins. For everything else, no problem. The first stage consists of the construction of the cube. For this, it is recommended to have a wooden board and make a template where the LEDs will go when they are to be welded.

The area of this template will depend on the length of the longest leg of your led. To open the holes, the use of a drill with a 5 mm bit is recommended, which is more or less the diameter of the LEDs.

Since you have the template, the cathodes of the LEDs are bent. In this project, we will work on each level as a common cathode. This means that all the cathodes of each LED must be united and facing inwards.

There are three ways to identify which is the cathode:

- It is the shortest led pin

- The side that is flat on the led

- It is connected to the largest piece inside the led

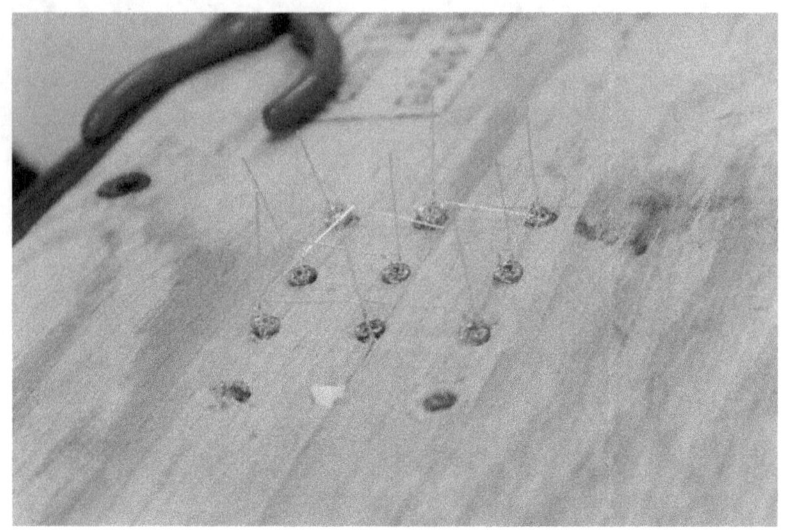

Use some threads to hold the common point between the two LEDs to be welded together. When you have finished soldering all the LEDs, and then repeat this procedure three times.

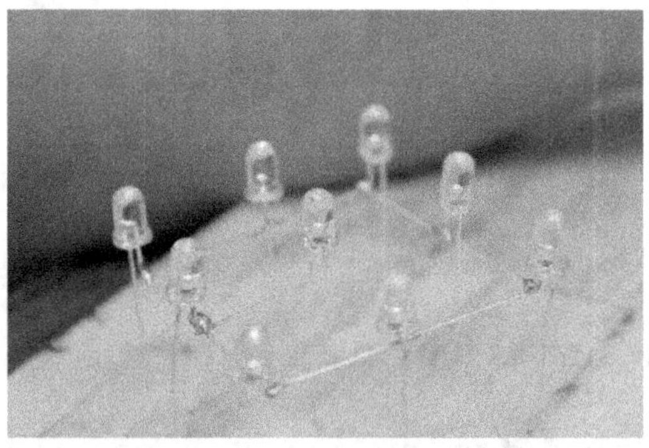

The next thing is to connect the levels, one on top of the other. For this, we will bend the anodes of the LEDs out a little, for when they are going to be welded, there is no problem.

We would have something like this.

The following is the circuit schematic:

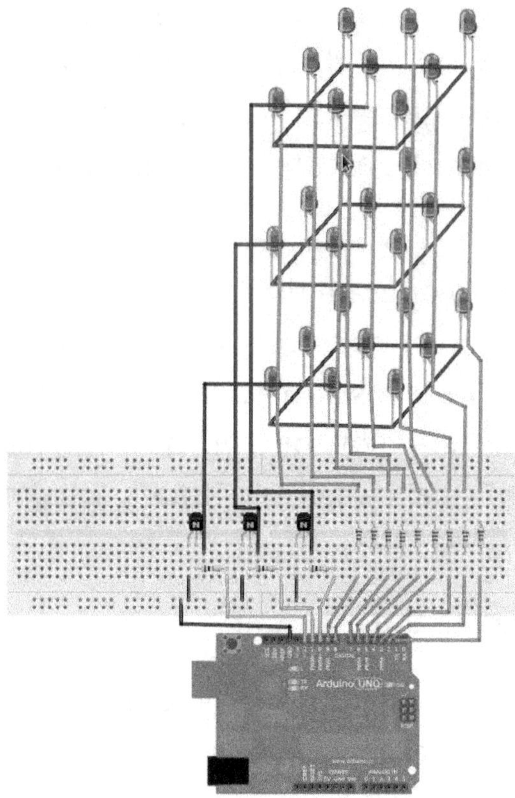

Now to connect the cube to the Arduino, we will need several components. One of them is resistors. As the Arduino provides each of its digital outputs with a voltage of 5 V, we cannot subject such a voltage to the LEDs. For that, a resistor is used that will decrease the applied voltage allowing our LEDs to not burn out. In this case, we use a value of 220 Ω.

Now, to make the LEDs turn on, we need to use the transistor in switching mode, that is, the transistors change their region of operation: from cut-off to saturation. So when the Arduino sends a small current (no more than 40 mA), the transistor activates in saturation causing the collector to "connect" with the emitter directing the cathodes to the ground.

In this case, to operate the transistor, switching it is necessary to connect the base of the transistor through a 10 kΩ resistor with the Arduino's digital pins, while the collector will be connected to the cathodes of the cube, and finally the emitter to ground.

Step 2: Creating the program

The only thing left to do is download the program created in processing/wiring for our Arduino UNO that will allow us to turn on the LEDs. Also, if you want, you can create your own LED lighting program.

There are a lot of examples that will allow us to make our cube, beauty in action =)

However, we can test the cube with the following code:

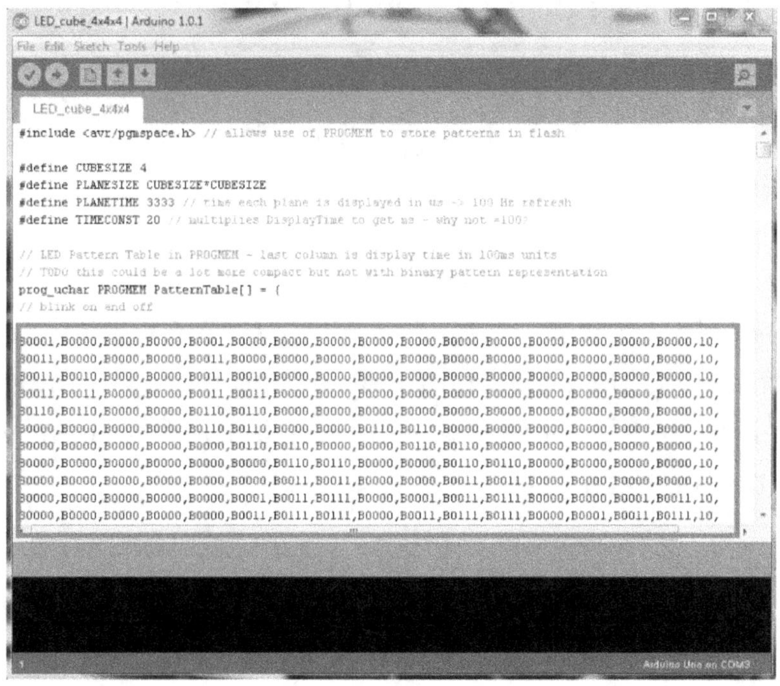

How to use Arduino libraries

To use the Arduino libraries, you must first unzip the library file you want to use, moving it to the

Arduino / sketchbook/libraries folder. If you are working, you can find this folder in the directory:

/ home / user

If you don't see the sketchbook folder, you can enter the Arduino IDE and in the File> Preferences tab you can find the address of the sketchbook directory. If you don't have the libraries folder created inside the sketchbook, create it.

Restart the Arduino IDE (if you had it open) and look in File> Sketchbook> libraries>LedCube> led cube.

Conclusion

As I said in the introduction, this beginner's guide to Arduino is designed for people who want to understand what is behind Arduino. It is the first contact with free hardware.

After this extensive tutorial, there would be a lot to do with programming, but with this, you already have a good foundation to build on.

Learning Arduino, like any other discipline, requires prior effort. You have to go through a learning curve. However, if you have a passion for electronics and programming, it seems even fun.

About the Author

Daniel Stone is dedicated to the world of programming and computer technology. He has written a number of books, which have met positive reviews. The beginners guide to Arduino Pro is another entry to his collection.

www.ingramcontent.com/pod-product-compliance
Lightning Source LLC
Chambersburg PA
CBHW052012150726
47999CB00004B/1630